C0-AMO-195

CRC SERIES IN AGING

Editors-in-Chief

Richard C. Adelman, Ph.D.   George S. Roth, Ph.D.

VOLUMES AND VOLUME EDITORS

HANDBOOK OF BIOCHEMISTRY IN AGING
James Florini, Ph.D.
Department of Biology
Syracuse University
Syracuse, New York

HANDBOOK OF IMMUNOLOGY IN AGING
Marguerite M. B. Kay, M.D. and Takashi Makinodan, Ph.D.
Geriatric Research Education and Clinical Center
V.A. Wadsworth Medical Center
Los Angeles, California

SENESCENCE IN PLANTS
Kenneth V. Thimann, Ph.D.
The Thimann Laboratories
University of California
Santa Cruz, California

ALCOHOLISM AND AGING: ADVANCES IN RESEARCH
W. Gibson Wood, Ph.D.
Clinical Research Psychologist
Geriatric Research, Education, and Clinical Center
V.A. Medical Center
St. Louis, Missouri
Merrill F. Elias, Ph.D.
Professor of Psychology
University of Maine at Orono
Orono, Maine

TESTING THE THEORIES OF AGING
Richard C. Adelman, Ph.D.
Director
Institute of Gerontology
Professor of Biological Chemistry
University of Michigan
Ann Arbor, Michigan
George S. Roth, Ph.D.
Research Biochemist
Gerontology Research Center
National Institute on Aging
Baltimore City Hospitals
Baltimore, Maryland

HANDBOOK OF PHYSIOLOGY IN AGING
Edward J. Masoro, Ph.D.
Department of Physiology
University of Texas Health Science Center
San Antonio, Texas

IMMUNOLOGICAL TECHNIQUES APPLIED TO AGING RESEARCH
William H. Adler, M.D. and Albert A. Nordin, Ph.D.
Gerontology Research Center
National Institute on Aging
Baltimore City Hospitals
Baltimore, Maryland

CURRENT TRENDS IN MORPHOLOGICAL TECHNIQUES
John E. Johnson, Jr., Ph.D.
Gerontology Research Center
National Institute on Aging
Baltimore City Hospitals
Baltimore, Maryland

NUTRITIONAL APPROACHES TO AGING RESEARCH
Gairdner B. Moment, Ph.D.
Professor Emeritus of Biology
Goucher College
Guest Scientist
Gerontology Research Center
National Institute on Aging
Baltimore, Maryland

ENDOCRINE AND NEUROENDOCRINE MECHANISMS OF AGING
Richard C. Adelman, Ph.D.
Director
Institute of Gerontology
Professor of Biological Chemistry
University of Michigan
Ann Arbor, Michigan
George S. Roth, Ph.D.
Research Biochemist
Gerontology Research Center
National Institute on Aging
Baltimore City Hospitals
Baltimore, Maryland

Additional topics to be covered in this series include Cell Biology of Aging, Microbiology of Aging, Pharmacology of Aging, Evolution and Genetics, Animal Models for Aging Research, Detection of Altered Proteins, Insect Models, and Lower Invertebrate Models.

Altered Proteins and Aging

Editors

Richard C. Adelman, Ph.D.
Director
Institute of Gerontology
Professor of Biological Chemistry
University of Michigan
Ann Arbor, Michigan

George S. Roth, Ph.D.
Research Chemist
Gerontology Research Center
National Institute on Aging
Baltimore City Hospitals
Baltimore, Maryland

CRC Series in Aging
Editors-in-Chief

Richard C. Adelman, Ph.D.
George S. Roth, Ph.D.

CRC Press, Inc.
Boca Raton, Florida

Library of Congress Cataloging in Publication Data
Main entry under title:

Altered proteins and aging.

(CRC series in aging)
Bibliography: p.
Includes index.
1. Proteins. 2. Aging. I. Adelman, Richard C.,
1940- . II. Roth, George S., 1946-
III. Series. [DNLM: 1. Proteins--Analysis.
2. Aging. WT 104 A466]
QP86.A473 1983 591.19'245 82-20619
ISBN 0-8493-5812-4

Direct all inquiries to CRC Press, Inc., 2000 Corporate Blvd., N.W., Boca Raton, Florida, 33431.

International Standard Book Number 0-8493-5812-4

Library of Congress Card Number 82-20619
Printed in the United States

FOREWORD

Analysis of protein structure and function has proved to be one of the most important challenging areas of biochemistry and molecular biology. The field is of no less significance to bio-gerontologists, since many investigators are convinced that changes in critical enzymes are at the least a key manifestation of aging. Indeed, whether or not one accepts a stochastic error theory of senescence, it must be conceded that many age-related functional changes are somehow associated with protein changes.

In this regard we have attempted to present here some of the more salient types of protein analysis with possible relevance to aging studies. This information should allow investigators to perform both critical analysis of past data and definitive future experiments in the area of proteins and aging.

George S. Roth
Richard C. Adelman

EDITORS-IN-CHIEF

Richard C. Adelman, Ph.D., is currently Director of the Institute of Gerontology at the University of Michigan, Ann Arbor, as well as Professor of Biological Chemistry in the Medical School. An active gerontologist for more than 10 years, he has achieved international prominence as a researcher, educator, and administrator. These accomplishments span a broad spectrum of activities ranging from the traditional disciplinary interests of the research biologist to the advocacy, implementation, and administration of multidisciplinary issues of public policy of concern to elderly people. He is the author and/or editor of more than 95 publications, including original research papers in refereed journals, review chapters, and books. His research efforts have been supported by grants from the National Institutes of Health for the past 13 consecutive years, and he continues to serve as an invited speaker at seminar programs, symposiums, and workshops all over the world. He is the recipient of the IntraScience Research Foundation Medalist Award, an annual research prize awarded by peer evaluation for major advances in newly emerging areas of the life sciences; and the recipient of an Established Investigatorship of the American Heart Association.

Dr. Adelman serves on the editorial boards of the *Journal of Gerontology, Mechanisms of Ageing and Development,* and *Gerontological Abstracts.* He chaired a subcommittee of the National Academy of Sciences Committee on Animal Models for Aging Research. As an active Fellow of the Gerontological Society, he was Chairman of the Biological Sciences section; a past Chairman of the Society Public Policy Committee; and is currently Chairman of the Research, Education and Practice Committee. He serves on National Advisory Committees which impact on diverse key issues dealing with the elderly, including a 4-year appointment as member of the NIH Study Section on Pathobiological Chemistry; the Executive Committee of the Health Resources Administration Project on publication of the recent edition of *Working with Older People — A Guide to Practice;* and a 4-year appointment on the Veterans Administration Advisory Council for Geriatrics and Gerontology.

George S. Roth, Ph.D., is a research chemist with the Gerontology Research Center of the National Institute on Aging in Baltimore, Md., where he has been affiliated since 1972. Dr. Roth received his B.S. in Biology from Villanova University in 1968 and his Ph.D. in Microbiology from Temple University School of Medicine in 1971. He received postdoctoral training in Biochemistry at the Fels Research Institute in Philadelphia, Pa. Dr. Roth has also been associated with the graduate schools of Georgetown University and George Washington University where he has sponsored two Ph.D. students.

He has published more than 70 papers in the area of aging and hormone/neurotransmitter action, and has lectured, organized meetings, and chaired sessions throughout the world on this subject.

Dr. Roth's other activities include fellowship in the Gerontological Society of America, where he has served in numerous capacities, including chairmanship of the 1979 midyear conference on "Functional Status and Aging." He is presently Chairman of the Biological Sciences Section and a Vice President of the Society. He has twice been selected as an exchange scientist by the National Academy of Sciences and in this capacity has established liaisons with gerontologists, endocrinologists, and biochemists in several Eastern European countries. Dr. Roth serves as an editor of *Neurobiology of Aging* and is a frequent reviewer for many other journals including *Mechanisms in Aging and Development, Life Sciences, The Journal of Gerontology, Science* and *Endocrinology.* He also serves as a grant reviewer for several funding agencies including the National Science Foundation. In 1981 Dr. Roth was awarded the Annual Research Award of the American Aging Association.

CONTRIBUTORS

Bruce J. Baum, D.M.D., Ph.D.
Dental Officer
Staff Scientist
Laboratory of Molecular Aging
National Institute on Aging
Baltimore, Maryland

Robert S. Bienkowski, Ph.D.
Assistant Professor of Pediatrics
Albert Einstein College of Medicine
Bronx, New York

Donald J. Cannon, Ph.D.
Department of Biochemistry
Veterans Administration Medical Center
University of Arkansas for Medical Sciences
Little Rock, Arkansas

Vernon J. Choy, Ph.D.
Research Fellow in Neurosciences
Department of Medicine
Flinders Medical Center
Bedford Park, Australia

Jean Claude Dreyfus, M.D., Ph.D.
Professor of Medical Biochemistry
Universite Paris
Director, French Medical Council
Paris, France

Robert W. Gracy, Ph.D.
Professor and Chairman
Department of Biochemistry
North Texas State University
Denton, Texas

Gerald P. Hirsch, Ph.D.
President
Dtec Company
Austin, Texas

Axel Kahn, M.D., D.Sc.
Maitre de Recherche a l'INSERM
Paris, France

Hsieng S. Lu, Ph.D.
Departments of Biochemistry and Chemistry
North Texas State University
Texas College of Osteopathic Medicine
Denton, Texas

Ian Phillips, Ph.D.
Department of Biochemistry
University College London
London, England

Morton Rothstein, Ph.D.
State University of New York
Division of Cell and Molecular Biology
Buffalo, New York

Fanny Schapira, M.D., Ph.D.
Director
Recherches au Centre National de la Recherche Scientifique
Paris, France

Anne Shephard, Ph.D.
Department of Biochemistry
University of College London
London, England

Gary S. Stein, Ph.D.
Professor
Department of Biochemistry and Molecular Biology
University of Florida
Gainesville, Florida

Janet L. Stein, Ph.D.
Assistant Professor
Department of Immunology and Medical Microbiology
University of Florida
Gainesville, Florida

CONTRIBUTORS

John M. Talent, Ph.D.
Department of Biochemistry and Chemistry
North Texas State University
Texas College of Osteopathic Medicine
Denton, Texas

Pau M. Yuan, Ph.D.
Departments of Biochemistry and Chemistry
North Texas State University
Texas College of Osteopathic Medicine
Denton, Texas

TABLE OF CONTENTS

Chapter 1
Detection of Altered Proteins 1
Morton Rothstein

Chapter 2
Structural Analysis of Altered Proteins 9
Robert W. Gracy, Hsieng S. Lu, Pau M. Yuan, and John M. Talent

Chapter 3
Error Measurement Methods in Aging Research 35
Gerald P. Hirsch

Chapter 4
Measurement of Intracellular Protein Degradation 55
Robert S. Bienkowski and Bruce J. Baum

Chapter 5
Approaches to Studying Age-Dependent Changes in Chromosomal Proteins 81
Ian Phillips, Anne Shephard, Janet L. Stein, and Gary Stein

Chapter 6
Molecular Mechanisms of Alterations of Some Enzymes in Aging 113
Jean Claude Dreyfus, Axel Kahn, and Fanny Schapira

Chapter 7
Altered Polypeptide Hormones and Aging 135
Vernon J. Choy

Chapter 8
Collagen and Aging 161
Donald J. Cannon

Index 169

Chapter 1

DETECTION OF ALTERED PROTEINS

Morton Rothstein

TABLE OF CONTENTS

I. Background 2

II. Initial Detection 2

III. Purification of Altered Enzymes 3

IV. Heat Sensitivity 4

V. Spectral Properties 5

VI. Other Procedures 5

VII. Properties Showing No Change 6

VIII. Precautions 6

IX. Summary 7

References 8

I. BACKGROUND

Preceding the search for structurally altered enzymes, there were sporadic attempts to determine whether various enzyme activities change with age. In this work, which has been summarized in two review articles,[1,2] enzyme assays were performed on homogenates of various rodent tissues. The reported differences in enzyme activity between young and old animals are generally small and rarely over 30%. They are often inconsistent and a number of the results are conflicting. The reasons would seem to lie in the use of a variety of animal strains maintained under differing conditions as well as measurement of specific activities of enzymes in crude preparations. A more recent study using animals grown under standardized environmental conditions showed no consistent age-related pattern in the levels of glutamic, malic, and lactic dehydrogenases.[3]

The search for altered enzymes in aging organisms was initiated by the error catastrophe theory of Orgel.[4] In brief, the theory proposes that because of an initial lesion (mutation), an altered protein would be formed. If this protein were part of the protein synthesizing system, errors would be manifested in newly synthesized proteins. If some of these proteins were themselves involved in the machinery of protein synthesis, they would cause even more errors in the next generation of proteins. The effect would escalate until an "error catastrophe" occurred. Though recent work has shown the theory to be implausible,[5-8] and indeed, Orgel himself early recognized this situation,[9] the hypothesis nonetheless, served a useful function in focusing attention on the effect of aging on protein structure through a search for altered enzymes.

II. INITIAL DETECTION

The detection of altered enzymes is based primarily upon finding enzymes with a reduced catalytic ability in aged animals. This characteristic must be determined by comparing samples of pure enzyme or by immunotitration. Simple measurements of enzyme activity in crude homogenates do not suffice, as a lowered activity per mg of protein may simply reflect the presence of less enzyme molecules, not damaged ones. For example, the specific activity of triosephosphate isomerase from aged *Turbatrix aceti,* a free-living nematode[10] showed the pattern expected of altered enzymes. That is, in crude homogenates, and in fact even after partial purification, the enzyme from old nematodes showed a specific activity of about 50% of that of the "young" enzyme. However, after the last step of purification (ion-exchange chromatography), the specific activities became identical. The reason for the apparent lower specific activity was that there was simply half as much triosephosphate isomerase in "old" homogenates, though the molecules were perfectly normal. Even the measurement of specific activity may turn out to be an inadequate criterion for altered enzymes. In the case of rat muscle phosphoglycerate kinase, specific activity is the same for young and old animals though respective enzymes are quite different in stability and show substantial spectral differences.

A second method for the detection of altered enzymes is based upon immunotitration. That is, determination of activity per unit of antiserum or the amount of antiserum required for 50% loss of activity. In this manner, altered enzymes with a reduced catalytic ability can be detected in crude homogenates because more molecules must be present to provide a given amount of activity. Therefore, more antiserum will be required. An immunotitration procedure was the basis of the original detection of altered isocitrate lyase in crude homogenates of aged *Turbatrix aceti.*[11] Use of antisera prepared to pure enzymes has shown that there is no change in the specific activity of

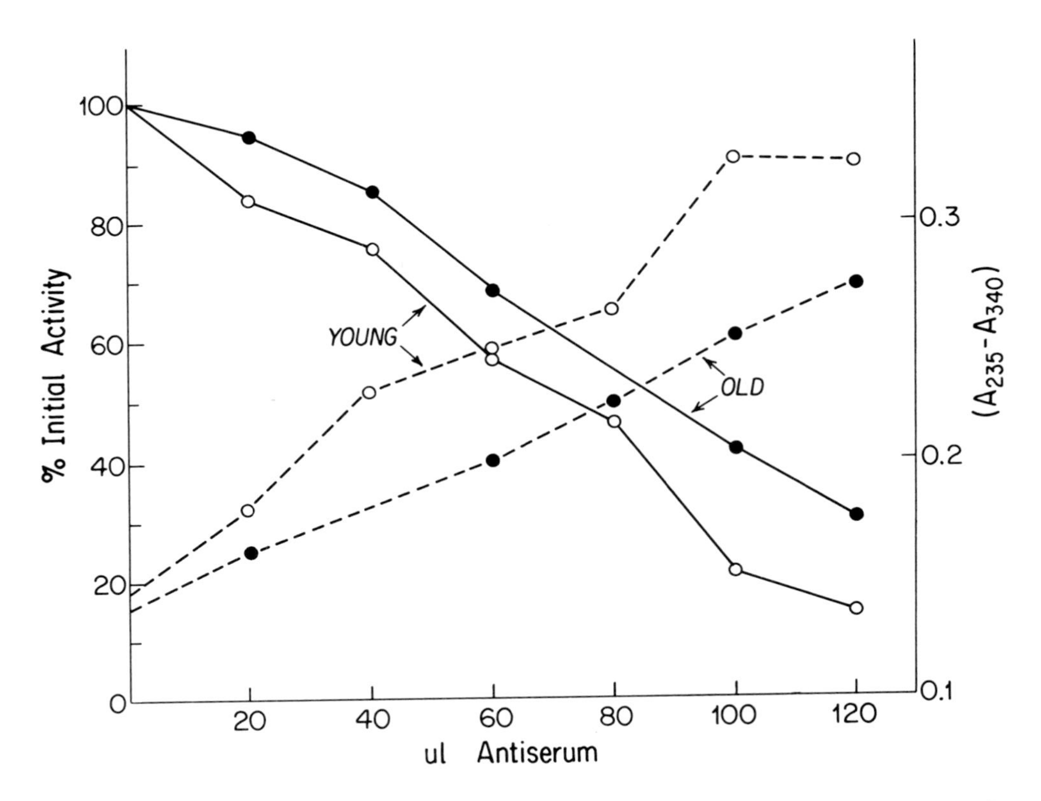

FIGURE 1. Titration of "young" and "old" enolase with antiserum prepared to "young" enolase. o——o, "young" enolase; ●——● "old" enolase, o- - - -o, ●- - - -●, A_{235}-A_{340} values for "young" and "old" enolase, respectively. The latter measurements, represent the amount of protein present in the immunoprecipitate. The amount of "young" protein precipitated is greater per unit of "young" antiserum, indicating a greater affinity of the antigen for its corresponding antiserum. Analogous relationships apply to Figure 2.

seven enzymes (pyruvate kinase, glucose-6-phosphate dehydrogenase, glucose phosphate isomerase, 6-phosphogluconate dehydrogenase, lactic dehydrogenase, α-mannosidase, β-glycuronidase) is extracts of granulocytes from aged humans.[12]

Antiserum prepared to "young" enzyme may react more effectively with "young" than with "old" enzyme and vice versa. Such is the case with enolase from *T. aceti*,[13] (Figures 1 and 2). No differences are observed on immunodiffusion plates.

III. PURIFICATION OF ALTERED ENZYMES

In general, the "old" enzymes isolated thus far, have not behaved differently from their "young" counterparts during purification. Location of enzyme peaks from columns, yield after various steps such as ammonium sulfate precipitation or chromatographic procedures and general behavior appear to be identical for "young" and "old" enzymes. In the case of "old" nematode enolase a small amount of denatured enzyme (detected by immunological procedures) precedes the peak of activity emerging from the DEAE-cellulose (DE-52) column.[14] A small amount of immunologically identical, inactive material was obtained from "young" enolase by passing the enzyme through the ion-exchange column a second and third time.[7] The "young" enzyme recovered after this treatment had become similar in properties to "old" enzyme.[7] These results have been interpreted to mean that "old" enolase is a slightly denatured form of "young" enolase.

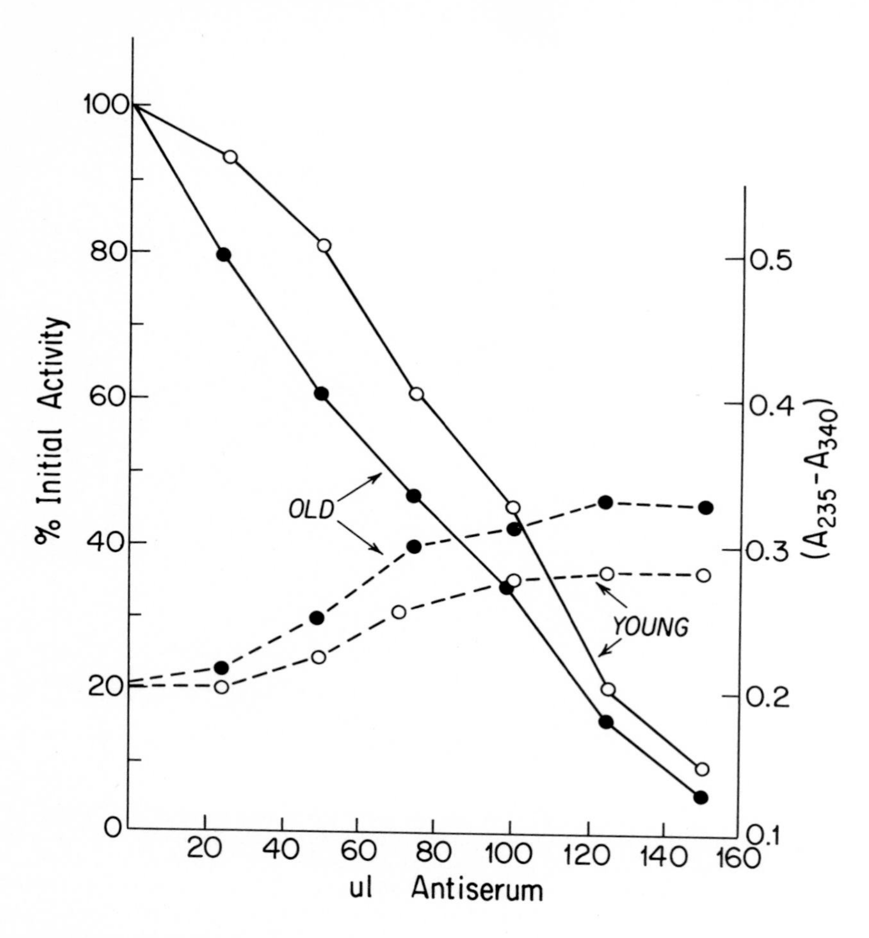

FIGURE 2. Titration of "young" and "old" enolase with antiserum prepared to "old" enolase. Symbols are the same as for Figure 1. In this case, "old" antiserum shows a greater reactivity with "old" enzyme than with "young" enzyme.

IV. HEAT SENSITIVITY

A secondary criterion for the detection of altered enzymes is a modified sensitivity to heat. Typically, "young" enzymes, when heated, yield a straight line relationship showing decreasing activity with time (A, Figure 3). Altered enzymes often, but not always, show a biphasic pattern, one component being heat-sensitive (B, Figure 3) and the second component either paralleling the "young" pattern (C, Figure 3) or remaining more sensitive to heat than the young enzyme (D, Figure 3). Thus, altered nematode isocitrate lyase[15] shows a pattern represented by BC, altered nematode enolase,[14] BD and altered rat liver superoxide dismutase,[16] BD (though the pattern is unusual in that the first component (B) is less heat-sensitive and the second component has the steeper slope). In the case of "young" and "old" nematode aldolase, a biphasic pattern is obtained for both enzymes.[17] For rat muscle phosphoglycerate kinase, straight lines are obtained for both "young" and "old" enzymes. Curiously, the latter is the more stable of the two forms. Nematode phosphoglycerate kinase yields an entirely different result as "young" and "old" (altered) enzymes show no difference in heat sensitivity.[18] In short, the results of temperature sensitivity measurements are unpredictable, but provide a useful though not infallible criterion for the detection of altered enzymes.

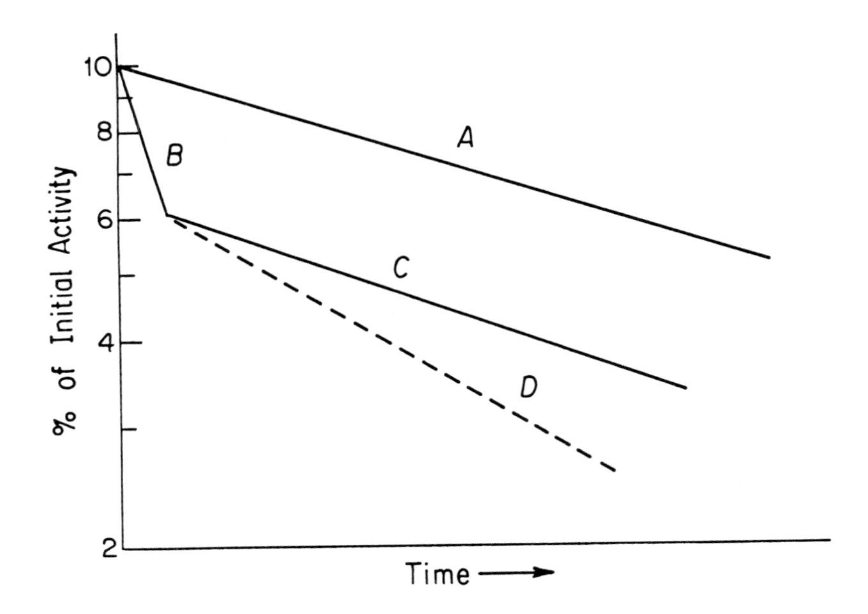

FIGURE 3. Heat-sensitivity of altered enzymes. A, B, C, D represent patterns of loss of activity at elevated temperatures.

As an adjunct to measuring heat stability, it should be noted the "young" and "old" enzymes often show differences in stability during storage.

V. SPECTRAL PROPERTIES

Spectral properties have thus far been compared only in two cases of altered enzymes: nematode enolase[7] and rat muscle phosphoglycerate kinase. In both cases, substantial differences were observed. Thus, such parameters as absorbance maximum and $A_{280/260}$ ratios may be valuable for confirming that an enzyme is altered. For "old" enolase the absorbance maximum and minimum were slightly shifted and the A_{280}/mg was substantially increased for "old" enolase (0.8 to 1.0 vs. 0.6). The $A_{280/260}$ ratio was substantially greater for the "young" enzyme (1.9 vs. 1.3). In the case of rat muscle phosphoglycerate kinase, the A_{280}/mg was 5.2 and 7.2 for "young" and "old" enzyme, respectively. Circular dichroism, which shows differences in enolase from young and old *T. aceti*,[7] should also be useful for showing changes in the structure of other altered enzymes.

For "young" and "old" enolase, all spectral differences disappeared when the molecules were unfolded in 6M guanidine. Other enzymes have not been tested in this respect.

VI. OTHER PROCEDURES

Any method for determination of protein conformation should be of value in studies of altered enzymes, particularly since current evidence strongly supports the idea that "old" enzymes are a result of conformational changes. For example, titration of "old" enolase showed two extra tyrosine hydroxyl groups and one extra tryptophan residue exposed to the medium,[7] though for each amino acid, the total number of residues was the same for both "young" and "old" enzyme. The titration results were confirmed by showing differences in C.D. spectra relating to these amino acids. α-Helical and random coil structure also showed differences.

Determination of SH groups is an obvious consideration since "old" enzymes might be considered to result from formation of S-S bonds. However, for both nematode enolase and rat muscle phosphoglycerate kinase, respectively, the number of SH groups was found to be identical for both young and old forms of the enzyme. Oxidation of methionine to the sulfoxide also may be considered as a possible cause of altered enzymes. Treatment of the suspected enzyme with high levels of sulfhydryl reagents (e.g. 0.1 *M* β-mercaptoethanol) should reduce any such oxidized products. Enzymes, if they do indeed become altered because of the formation of methionine sulfoxide, should be normalized, or at least, should show changed properties after this treatment. No such effects were observed with "old" enolase.

VII. PROPERTIES SHOWING NO CHANGE

None of the altered enzymes so far discovered show differences with respect to polyacrylamide gel electrophoresis or isoelectric focusing. Km values show minor variations which may or may not be significant. Where tested, inhibitor studies have not demonstrated kinetic differences. These techniques, then, are of little value in detecting or characterizing altered enzymes. Affinity chromatography of "young" vs. "old" isocitrate lyase[15] also showed no differences in elution pattern. This result would perhaps be expected from the lack of difference in Km values. Immunodiffusion plates also fail to show differences between "young" and "old" enzymes. However, a form of enolase designated "inactive enolase" was detected in homogenates of old *T. aceti,* and from chromatographic columns during purification of the "old" enzyme.[7,14] This material showed a spur when tested against antiserum prepared to the "old" enzyme.[13] It is presumed that "inactive enolase" is a denatured product of "old" enolase.

Amino acid analysis has not proved useful in that small errors in sequence, if they occur, would not be detectable. Moreover, the evidence available strongly suggests that the alteration of enzymes is not due to this cause.

VIII. PRECAUTIONS

Precautions should be taken to guarantee that the appearance of altered enzymes is not a result of some component specifically present in old tissues which interacts with the enzyme during its isolation. In this respect, the most obvious fear is of differential proteolytic action. For example, one might postulate that in "old" tissues, increased levels of lysosomal proteases are released during isolation of an enzyme so that proteolytic action brings about alteration of its properties. In this regard a recent report provides evidence that if "old" rat liver homogenates are stored, the C-terminal tyrosine of aldolase is lost, resulting in "altered" aldolase,[19] No such change occurs in "young" homogenates, nor is aldolase altered when isolated from fresh "old" preparations. These results may explain why Gershon and Gershon[20] obtained altered aldolase in old mouse liver (the authors stored their homogenates) but Anderson[21] and Weber et al.[22] reported no age-related change in rat liver aldolase.

One should bear in mind that the above results apply only to stored liver homogenates and specifically to aldolase. In our hands, frozen storage of *intact* rat muscle and liver tissue (up to 2 years) has no effect on young or old enolase or phosphoglycerate kinase, There is also the constraint that during proteolysis, no amino acid residues be lost so as to change the net charge of the protein; for each young-old enzyme pair studied (enolase,[14] isocitrate lyase,[15] superoxide dismutase[16]) isoelectric focusing showed no difference in charge. This point was further emphasized in a recent isoelectric focusing study of the latter two enzymes.[23] Nonetheless, to obtain unequivocal results, it is mandatory to show that the respective C- and N-terminal amino acids of putative altered enzymes are identical with those of their normal counterparts. Such experiments have so far been done for altered nematode enolase[7] and altered rat muscle phosphoglycerate kinase. Experiments showing that addition of inhibitors of proteolysis or mixing of young and old tissues have no effect on the enzymes add considerably to confidence though one may argue that they are not unequivocal.

There are other dangers inherent in studying altered enzymes in crude homogenates. Already mentioned is the fact that measurement of specific activity under these conditions is meaningless as one cannot distinguish between the presence of fewer molecules and catalytically less effective ones. Studies of heat sensitivity can also provide erroneous results. For example, it was reported that in crude homogenates, "young" and "old" nematode aldolase showed no difference in this parameter,[24] though the latter was an altered enzyme. Yet, when nematode aldolase was later purified, the heat sensitivity of the "old" enzyme was found to differ substantially from that of the "young" enzyme, actually being more stable.[17] When the heat sensitivity of glucose-6-phosphate dehydrogenase (G6PD) was determined in a variety of crude young and old mouse tissues,[25] the slopes of the curves varied substantially from sample to sample though the enzyme should be the same. Schofield and Hadfield[26] point out that 6-phosphogluconate dehydrogenase contributes significantly to the activity as measured by the assay used in the above system. Holliday and Tarrant studied G6PD in homogenates of early and late passage cells.[27] Examination of the figures shows that the rate of loss of the enzyme during heating of comparable cells varied in different experiments. In fact, Kahn et al.[28] showed that enzymes may be altered by components in cell homogenates. Pure "young" (unaltered) G6PD, added to homogenates of late-passage cells from which "old" enzyme had been removed by precipitation with antiserum, became "altered" with respect to heat sensitivity.

The presence of isozymes which may change in amount with age may also be a problem when working with crude enzyme preparation. In *T. aceti,* the ratios of the isozymes of isocitrate lyase changed with age.[15,29] Age-related changes in the amounts of hexameric, tetrameric, and dimeric forms of G6PD have been reported in rat liver.[30]

Another precaution to be considered is the use of experimental animals at as old an age as practical. At the present time, it is unknown at what age in rats a given enzyme will become altered, if alteration occurs at all.

IX. SUMMARY

Each altered (old) enzyme shows a unique set of differences when compared to its unaltered (young) counterpart. In general, a reduced specific activity is the simplest characteristic to detect. Once purified, other changes will come to light which may involve differences in spectral and immunological properties and altered sensitivity to heat or storage. It is unlikely that Km values, inhibitor studies, polyacrylamide gel electrophoresis, isoelectric focusing, or immunodiffusion will show differences. Care should be taken to avoid artifacts such as may occur in crude or partially purified preparations. The possibility that proteolytic action occurs during isolation should be eliminated by showing that C- and N-terminal groups are unchanged in the young-old enzyme pairs being compared.

REFERENCES

1. Finch, C. E., Enzyme activities, gene function and ageing in mammals (review), *Exp. Gerontol.*, 7, 53, 1972.
2. Wilson, P. D., Enzyme changes in ageing mammals, *Gerontologia*, 19, 79, 1973.
3. Kerr, J. S. and Frankel, H. M., Substrate, enzyme and coenzyme levels in male and female rats from 3 to 24 months of age, *Int. J. Biochem.*, 7, 455, 1976.
4. Orgel, L. E., The maintenance of the accuracy of protein synthesis and its relevance to aging, *Proc. Natl. Acad. Sci. USA*, 49, 517, 1963.
5. Rothstein, M., Recent developments in the age-related alteration of enzymes, a review, *Mech. Age. Dev.*, 6, 241, 1977.
6. Rothstein, M., The formation of altered enzymes in aging animals, *Mech. Age. Dev.*, 9, 197, 1979.
7. Sharma, H. K. and Rothstein, M., Age-related changes in the properties of enolase from *Turbatrix aceti*, *Biochemistry*, 17, 2869, 1978.
8. Edelman, P. and Gallant, J., On the translational error theory of aging, *Proc. Natl. Acad. Sci., USA*, 74, 3396, 1977.
9. Orgel, L. E., Ageing of clones of mammalian cells, *Nature (London)*, 243, 441, 1973.
10. Gupta, S. K. and Rothstein, M., Triosephosphate isomerase from young and old *Turbatrix aceti*, *Arch. Biochem. Biophys.* 174, 333, 1976.
11. Gershon, H. and Gershon, D., Detection of inactive enzyme molecules in aging organisms, *Nature (London)*, 227, 1214, 1970.
12. Rubinson, H., Kahn, A., Boivin, P., Schapira F., Gregori, C. and Dreyfus, J. C., Aging and accuracy of protein synthesis in man: search for inactive enzymatic cross-reacting material in granulocytes of aged people, *Gerontology*, 22, 438, 1976.
13. Sharma, H. K. and Rothstein, M., Serological evidence for the alteration of enolase during aging, *Mech. Age. Dev.*, 8, 341, 1978.
14. Sharma, H. K., Gupta, S. K., and Rothstein, M., Age-related alteration of enolase in the free-living nematode *Turbatrix aceti*, *Arch. Biochem. Biophys.*, 174, 324, 1976.
15. Reiss, U. and Rothstein, M., Age-related changes in isocitrate lyase from the free-living nematode, *Turbatrix aceti*, *J. Biol. Chem.*, 250, 826, 1975.
16. Reiss, U. and Gershon, D., Rat liver superoxide dismutase: purification and age-related modifications, *Eur. J. Biochem.*, 63, 617, 1976.
17. Reznick, A. Z. and Gershon, D., Age-related alterations in purified fructose-1,6-diphosphate aldolase from the nematode, *Turbatrix aceti*, *Mech. Age. Dev.*, 6, 345, 1977.
18. Gupta, S. K. and Rothstein, M., Phosphoglycerate kinase from young and old *Turbatrix aceti*, *Biochim. Biophys. Acta*, 445, 632, 1976.
19. Petell, J. K. and Lebherz, H. G., Properties and metabolism of fructose diphosphate aldolase in livers of "old" and "young" mice, *J. Biol. Chem.*, 254, 8179, 1979.
20. Gershon, H. and Gershon, D., Inactive enzyme molecules in aging mice: liver aldolase, *Proc. Natl. Acad. Sci. USA*, 70, 909, 1973.
21. Anderson, P. J., The specific activity of aldolase in the livers of old and young rats, *Can. J. Biochem.*, 54, 194, 1976.
22. Weber, A., Gregori, C., and Schapira, F., Aldolase B in the liver of senescent rats, *Biochim. Biophys. Acta*, 444, 810, 1976.
23. Goren, P., Reznick, A. Z., Reiss, U., and Gershon, D. Isoelectric properties of nematode aldolase and rat liver superoxide dismutase from young and old animals, *FEBS Lett.*, 84, 83, 1977.
24. Zeelon, P., Gershon, H., and Gershon, D., Inactive enzyme molecules in aging organisms. Nematode fructose-1,6-diphosphate aldolase, *Biochemistry*, 12, 1743, 1973.
25. Wulf, J. H. and Cutler, R. G., Altered protein hypothesis of mammalian ageing processes. I. Thermal stability of glucose-6-phosphate dehydrogenase in C57BL/6J mouse tissue, *Exp. Gerontol.*, 10, 101, 1975.
26. Schofield, J. D. and Hadfield, J. M., Age-related alterations in the heat-lability of mouse liver glucose-6-phosphate dehydrogenase, *Exp. Gerontol.*, 13, 147, 1978.
27. Holliday, R. and Tarrant, G. M., Altered enzymes in ageing human fibroblasts, *Nature (London)*, 238, 26, 1972.
28. Kahn, A., Gillouzo, A., Leibovitch, M. P., Cottreau, P., Bourel, M., and Dreyfus, J. C., Heat lability of glucose-6-phosphate dehydrogenase in some senescent human cultured cells. Evidence for its postsynthetic nature, *Biochem. Biophys. Res. Comm.*, 77, 760, 1977.
29. Reiss, U. and Rothstein, M., Heat-labile isozymes of isocitrate lyase from aging *Turbatrix aceti*, *Biochem. Biophys. Res. Comm.*, 61, 1012, 1974.
30. Wang, R. K. J. and Mays, L. L., Isozymes of glucose-6-phosphate dehydrogenase in livers of aging rats, *Age*, 1, 2, 1978.

Chapter 2

STRUCTURAL ANALYSIS OF ALTERED PROTEINS

Robert W. Gracy, Hsieng S. Lu, Pau M. Yuan, and John M. Talent

TABLE OF CONTENTS

I. Introduction 10

II. Preliminary Considerations 11
- A. Reagent Purity 11
- B. Glassware 11
- C. Proteases 11

III. High Sensitivity Homology Structural Analysis 12
- A. Rationale 12
- B. Primary Fragmentation 12
- C. Separation of Peptides 13
- D. Recovery of Peptides from SDS-Polyacrylamide Gels 16
- E. Peptide Alignment 17
- F. Homology Peptide Mapping 17
  - 1. Fragmentation Into Small Peptides 17
  - 2. Electrophoresis 20
  - 3. Chromatography 21
  - 4. Staining and Visualization of Peptides 21
  - 5. Recovery of Peptides from Thin-Layers 21
- G. High Sensitivity Amino Acid Analysis 21
- H. Amino Terminal Determination 22
- I. Carboxyl Terminal Determination 23
- J. Microsequencing 24
  - 1. General Methods 24
  - 2. Manual Sequencing Methods 25
    - a. Microsequencing Procedure Using Back Hydrolysis 25
    - b. Microsequencing Using *N,N*-Dimethylaminoazobenzene 4′-Isothiocyanate (DABITC) 26
  - 3. Detection of Phenylthiohydantoin Amino Acids by High Performance Liquid Chromatography 26

IV. Molecular Basis for Abnormal Triosephosphate Isomerase During Aging 27

Acknowledgment 28

References 31

I. INTRODUCTION

It has become increasingly clear from both in vitro and in vivo studies at the cellular, organ, and organism levels that "abnormal" proteins accumulate during aging. Although the relationships of these altered proteins to the overall biology of aging is unclear, it is likely that the accumulation of abnormal proteins could impair cells, tissues, etc., in responding to environmental changes. Numerous studies have reported "defective" enzymes and postulated alterations in the ability of old cells to synthesize or degrade proteins.[1-17] Enzyme activity, thermal stability, electrophoretic, kinetic, and immunological properties have been among the most common parameters measured in these studies (for reviews see References 18, 19, and other chapters of this series.) Considerable debate has resulted regarding the biochemical basis for these "abnormal" proteins in aging. Orgel[20] and Holliday et al.[1] proposed that a loss of fidelity in protein synthesis could lead to the accumulation of defective enzymes such that the cell was no longer capable of conducting normal physiological functions. Others[6,7,21-23] suggested that the changes are the consequences of postsynthetic protein modifications. Thirdly, the altered proteins could be the result of changes in the catabolism of proteins.[16,17,24-29]

The major reason that this question has not been resolved is that the differences between the *structures* of the "aged" proteins and their "normal" counterparts have not been established. If these differences were known it is almost certain that the mechanisms for the formation of abnormal proteins in aging would be much more apparent. The primary problem in elucidating these differences has been the difficulty in conducting detailed protein structural studies with the exceedingly small amounts of protein available in most aging studies. This is the case when dealing with aging animals, cells grown in tissue culture, and with aging human populations. Often even amounts of the "normal" proteins from the young controls are also quite limited. Moreover, the abnormal proteins must be isolated to homogeneity from both the young and aged groups prior to the initiation of structural analyses. Thus, isolation methods must be very selective and result in high recovery of the protein. Finally, it is likely that the abnormal proteins which accumulate with aging are the consequences of rather subtle structural changes. Therefore, the methods of analysis must be capable of detecting single amino acid changes and/or changes in the secondary, tertiary, and quarternary structures of these molecules.

Fortunately, methods of protein structural analysis have recently undergone advances resulting in marked increases in both analytical sensitivity and specificity. Methods such as affinity chromatography, high sensitivity amino acid analysis, and microsequencing now make it possible to determine complete structures with only a few micrograms of protein. For example, the amount of material recovered from a single polyacrylamide gel may be sufficient for complete sequence determination.[30-32] With the application of these methods and strategies as homology peptide mapping[33,34] it is now possible to explain age-related protein changes at the molecular level. For example, it has been possible to establish the primary structure and to construct a three-dimensional model of human triosephosphate isomerase with less than 5 mg of protein.[34] These studies have revealed the details of the molecular changes responsible for the acidic forms of the enzyme which accumulate during aging.[35]

The purpose of this chapter is to present a rationale, strategy, and detailed experimental design for ultrasensitive protein structural analyses. Human triosephosphate isomerase is an example of how such analyses can provide important clues to the mechanisms of formation of altered proteins during aging.

II. PRELIMINARY CONSIDERATIONS

A. Reagent Purity

In microanalytical analyses the purity of reagents is of utmost importance. For example, unless solvents are known to be amine-free, it is necessary to redistill them. A small amount of ninhydrin is added to the solvent to react with free amines and to render them nonvolatile during distillation. The solvent should then be distilled again after adding KOH pellets to prevent carryover of volatile acids. Water should be 65 megohms or better, and should be distilled just prior to use. Reagents frequently need to be recrystallized before they are suitable for use. Tilley et al.[36] have shown that commercial cellulose thin-layer plates are contaminated with significant levels of proteins which are not removed by washing and which can interfere with amino acid analyses at the nanomole and subnanomole levels. This necessitates the use of special methods of extraction of peptides from thin-layer plates. The levels of protein contamination vary greatly depending on supplier and lot number. Generally, a satisfactory cellulose thin-layer is the "Polygram Cel 1400" supplied by Machery and Nagal. It is essential to check a series of plates by analyzing control areas and to purchase sufficient quantities of a single lot. For high sensitivity amino acid analysis at the subnanomole level, it is best to purchase pre-made buffers. For example, the Dionex "Hi-Phi Eluent Buffer System" offers good, single column resolution and relatively noise-free backgrounds. All buffers used in microanalysis should be maintained under a nitrogen atmosphere.

B. Glassware

The cleanliness of glassware is crucial. When analyzing submicrogram quantities of proteins and peptides, contaminants may be present in quantities greater than the sample if special precautions are not taken. It is well known that amino acids and peptides from the hands can be a major source of contamination, thus disposable gloves should be worn at all stages. Trace contaminants left on reusable glassware may also be a major source of problems. A suitable cleaning procedure is described below. First glassware is scrubbed vigorously, rinsed and soaked at 50° for 4 to 6 hr in a freshly prepared solution of alkaline detergent (e.g., 2% RBS-35; Pierce Chemical Co.). Glassware is scrubbed again, rinsed thoroughly with distilled water, then soaked overnight in a freshly prepared acid-cleaning solution (e.g., HNO_3:-H_2SO_4, v/v, 1:1). After soaking overnight in the acid, the glassware is removed and rinsed thoroughly with distilled water. Glassware intended for peptide analysis is then immersed for several seconds in freshly prepared 1% Prosil-28 (PCR) or some other suitable siliconizing agent, rinsed thoroughly and dried by heating at 100° for 10 min. Acid-washed, air-dried glassware absorbs ammonia upon standing, and washing with water is inadequate to remove surface ammonia. Consequently, glassware intended for amino acid analysis is kept in an acid-filled desiccator until needed, dried in an oven, and used as soon as dry.

Several companies offer specialized glassware useful for high sensitivity protein analysis. One of the most useful of these is the thick-walled vials having conical inner wells (Pierce, Regis, Wheaton). The conical bottom permits concentration of the sample by drying or centrifugation, and they are available in sizes from 100 $\mu\ell$ to 5 mℓ with threaded tops that accept regular lids or septa. Magnetic stirring bars, Teflon® valves, and heating stirring modules are commercially available to fit these vials.

C. Proteases

Proteolytic modification of proteins is always a potential problem in structural studies. Proteolytic activity in cell homogenates can result in the erroneous interpretation of modifications believed to be due to aging. For example Petell and Lebherz[37] showed that the previously reported defective forms of aldolase found in "aged" mouse liver are due to proteolytic activity during storage of cell extracts. The addition of proteolytic inhibitors such as phenylmethylsulfonyl fluoride to the homogenizing buffer can prevent the formation of these altered proteins.[37-39] However, of the numerous claims of defective proteins in aging cells, very few of the studies have made use of protease inhibitors.[37] Conditions such as freezing, which result in lysozomal rupture can add to the potential for proteolytic modification in vitro. For example, freezing of liver prior to homogenization caused proteolytic modification of the carboxyl-terminus of aldolase.[39] Thus cells should be collected and the protein isolated as rapidly as possible and under conditions which will minimize the formation of proteolytic artifacts.

III. HIGH SENSITIVITY HOMOLOGY STRUCTURAL ANALYSIS

A. Rationale

In view of the high degree of sequence homology between proteins from different species, short cuts can be used to establish the primary structure of proteins which are available in extremely limited quantities by simultaneously analyzing homologous "reference" proteins. For example, by utilizing simultaneous peptide mapping of rabbit and human triosephosphate isomerase it has been possible to establish the amino acid sequence of human triosephosphate isomerase with less than 5 mg of enzyme.

In order to conduct homology structural analysis, a source of the protein should be available which provides a substantial amount of "reference" protein (10 to 100 mg). It is also desirable that as much structural information as possible already be available on the reference protein. In aging studies, homology analysis is useful when one has available larger amounts of the protein from the young (control) group but very limited quantities from the aging group. In all cases the strategy is the same: protein cleavage, peptide separation, amino acid analysis, and sequencing are conducted simultaneously on the "rare" protein along with the more abundant homologous "reference" protein.

B. Primary Fragmentation

For sequence determination of a large protein, preliminary fragmentation to obtain a limited number of peptides is usually required. The use of a successful fragmentation procedure greatly simplifies isolation and alignment of the peptides. Typical fragmentation strategies have been reviewed elsewhere.[41,42] The initial fragmentation procedure should be very specific (i.e., only hydrolyzing the peptide bond between two specific types of amino acids), and the reaction should be quantitative (i.e., all of the molecules should be hydrolyzed to the same extent). In order to obtain a small number of unique peptides, it is advantageous to cleave the protein at amino acids or pairs of amino acids which occur at a low frequency within the molecule. Table 1 summarizes some fragmentation methods for obtaining a limited number of peptides. Two of the most widely utilized techniques involve cleavage of methionyl bonds with cyanogen bromide[43] and specific cleavage at argininyl bonds of ε-*N*-acylated proteins with trypsin.[44] Many reagents have been utilized for hydrolysis at tryptophanyl residues,[45-48] however, most of them result in incomplete cleavage or additional modification of other residues.

When initiating structural analysis of a new protein it is useful to test the susceptibility of the protein to acid cleavage of Asp-Pro bonds[49] or to Asn-Gly cleavage with hydroxylamine.[50] These procedures can be monitored conveniently by SDS-gel electrophoresis. This approach has been used for the cleavage of phosphorylase.[51] Similarly, human glucosephosphate isomerase (ca. 550 residues/subunit) was found to contain only three Asn-Gly bonds susceptible to hydroxylamine.[52] Hence, the rare occurence

Table 1
SPECIFIC LIMITED CLEAVAGE OF PROTEINS

Susceptible peptide bond	Cleavage agent	Ref.
Met-x	Cyanogen bromide (CNBr)	43
Trp-x	*N*-Bromosuccinimide	40
Trp-x	Bromo-*o*-nitrophenylsulfenyl (BNPS)-skatole	45
Trp-x	CNBr/anhydrous heptafluorobutyric acid	46
Trp-x	Iodosobenzoic acid	47
Trp-x	Dimethyl sulfoxide-hydrobromic acid	48
Asn-Gly	NH_2OH (pH 9.0)	50
Asp-Pro	Acid	49
x-Cys	2-Nitro-5-thiocyanobenzoic acid	54,55
Arg-x	Trypsin + Lys blocked peptide	44
Lys-x	Trypsin + Arg blocked peptide	58
Arg-x	Clostripain	56
Glu-x	Staphylococcus protease V8	57

of Asp-Pro and Asn-Gly sequences (approximately 1 per 200 residues) makes them ideal for this purpose. These methods can be successful even if the cleavage is not quantitative, and the size of overlap peptides can be useful in establishing the linear alignment of peptides (Figure 1).

Frequently the amino acid composition of the protein may suggest potential cleavage loci of suitable low frequency. For instance, the low sulfhydryl content of most proteins makes the reagent 2-nitro-5-thiocyanobenzoic acid (NTCB) particularly valuable for cleavage of cysteinyl peptide bonds. Guinea pig hemoglobin β-chain has been cleaved at the single cysteine by NTCB into two peptides.[53] Similarly, NTCB promotes cleavage of three cysteinyl bonds of human glucosephosphate isomerase in nearly quantitative yield as seen in Figure 2. When this method was initially utilized,[54,55] it appeared that the resulting blocked amino-termini prevented further Edman degradation. However, deblocking of the thiazolidine ring with Raney nickel results in a new amino-terminal alanine.[53]

The use of clostripain (argininyl bonds) and staphylococcal protease (glutamyl bonds) provide enzymatic methods for limited cleavage.[56,57] Modification of arginine residues with cyclohexadione[58] or butadione followed by treatment with trypsin results in specific cleavage at lysine.

One of the most selective protein hydrolysis methods is seldom used. The technique takes advantage of the intrinsic resistance of the compact structure of a native protein to proteolytic hydrolysis. For example, if a protein contains two distinct highly structured domains in its tertiary and quaternary structure, it may be possible to "clip" the protein at the juncture between the domains. In general, proteases with broad specificity, such as subtilisin, act at limited, specific positions if the time of treatment is controlled and the native structure of the substrate is preserved. For example, γ-globulin can be cleaved to two complementary chains by papain[59] or trypsin,[60] and ribonuclease[61] and phosphorylase[62] can be cleaved into two segments each by subtilisin. However, many proteins are degraded by proteases in a "one by one" model[63] in which the first hydrolysis produces an overall destabilization of the protein structure followed by a fast and extensive hydrolysis.

C. Separation of Peptides

The tactics of limited cleavage lead to relatively simple mixtures of a small number of large peptides. Although powerful methods have been developed for the isolation of proteins or mixtures of small peptides, neither of these methods nor their combina-

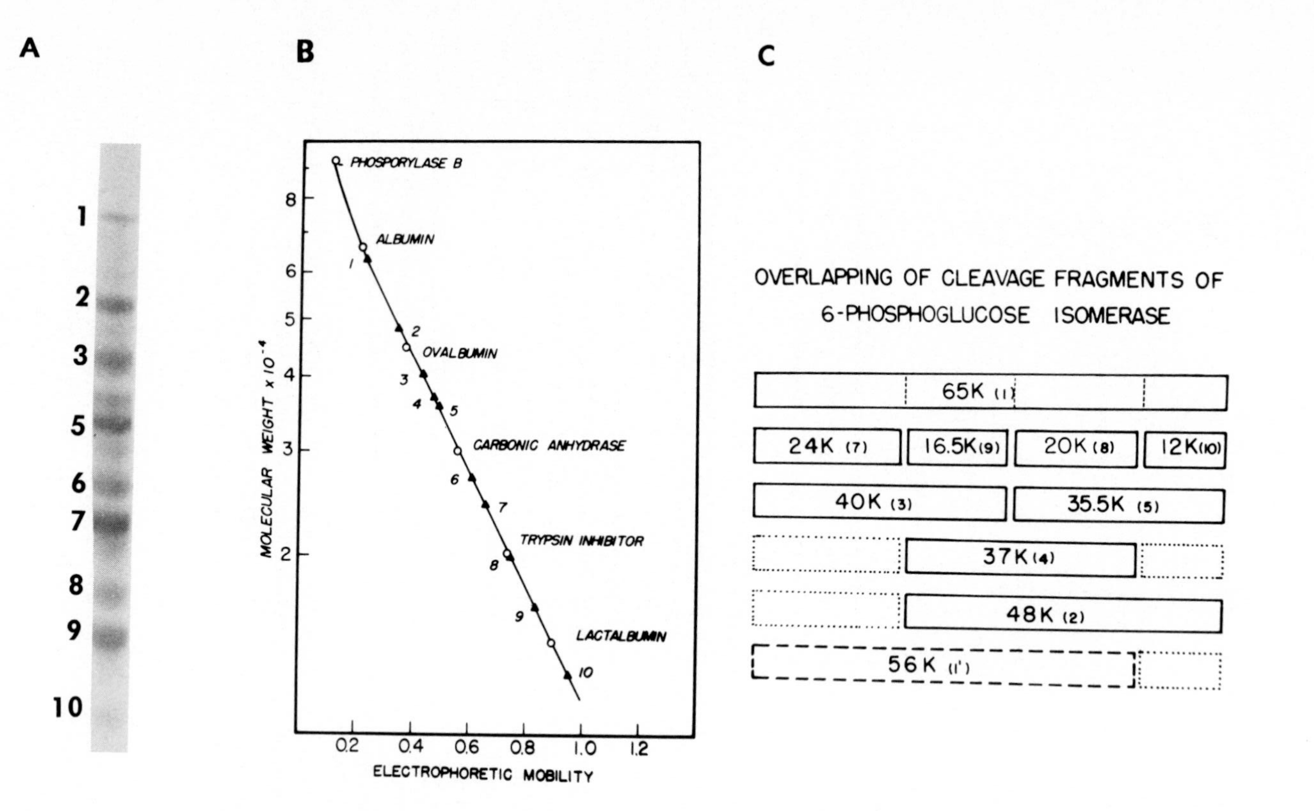

FIGURE 1. Limited cleavage of human glucosephosphate isomerase by hydroxylamine. (A) Polyacrylamide SDS-gels of the native enzyme (left) and the resulting peptides following cleavage at Asn-Gly bonds with hydroxylamine (see text for details). (B) Plot of migration of unknown peptides from human glucosephosphate isomerase (▲) and standard proteins (O) of known subunit molecular weight. (C) Reassembly of the primary peptides and overlapping fragments into the overall structure of the protein. The size of each peptide is indicated as (xK), and the number of each peptide refers to its migration in Figure 1A. The dashed-peptides represent those not observed on the gels.

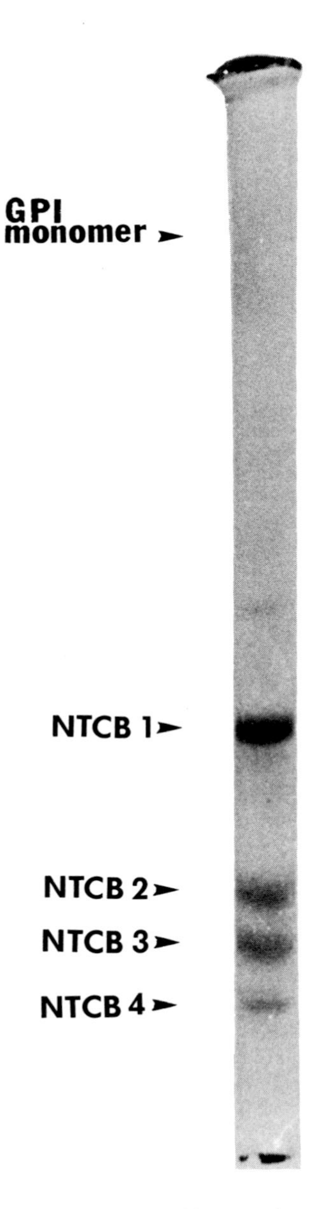

FIGURE 2. Cleavage of human glucosephosphate isomerase at cysteine residues with NTCB. The photography shows a 10% SDS-polyacrylamide gel of the enzyme which has been chemically cleaved at the three cysteines as described in the text. The gel was stained for total protein with Coomassie® blue. The markers show the position of the native subunit (GPI-monomer) and the four resulting peptides. NTCB-4 comigrates with tracking dye and shows diffusing banding.

tion is entirely suitable for mixtures of large polypeptides which tend to aggregate and precipitate. However, these problems can usually be overcome by polyacrylamide gel electrophoresis in the presence of sodium dodecyl sulfate, by isoelectric focusing in urea, or by gel filtration in denaturing solvents.

SDS polyacrylamide electrophoresis takes advantage of the excellent solubility of most polypeptides in SDS-solutions and the fact that the molecular weight of the polypeptide determines its mobility. Many gel systems are applicable for the separation of peptides, and SDS gel electrophoresis methods have been extensively reviewed by Maizel[64] and Weber and Osborn.[65] The size of SDS-polypeptide complex is proportional to the polypeptide chain length (i.e., molecular weight). Figure 1B shows an empirical calibration obtained in a SDS-gel electrophoresis experiment of hydroxylamine cleavage products of human glucosephosphate isomerase along with standards of known molecular weight.

Gel filtration is generally carried out in denaturing solvents at a pH as far removed as possible from the isoelectric point. High concentrations of formic or acetic acids are often suitable, but acidic peptides are generally more soluble in alkaline solutions of ammonium hydroxide. Urea or guanidinium chloride are alternative solutions but urea can lead to carbamylation of amino group at high pH and guanidinium chloride precludes the use of ion-exchange chromatography or isoelectric focusing.

Besides separation on the basis of size, peptides can be separated by ion-exchange chromatography in the presence of a nonionic caotropic agent. For example, β-galactosidase peptides were separated on carboxymethyl cellulose in 8 *M* urea,[66] and phosphorylase fragments were separated on SP-Sephadex in urea.[42] Similarly, isoelectric focusing in urea can provide high resolution of peptides. The use of high-pressure liquid chromatography, either by absorption or molecular sieving[67,68] is also an excellent choice for the isolation of primary peptide fragments.

D. Recovery of Peptides from SDS-Polyacrylamide Gels[69]

After separation of primary cleavage peptides by SDS-gel electrophoresis, the peptides must be recovered and subjected to structural analysis. The extraction method must be effective in obtaining high yields of the peptide, and, if the peptide to be stained, the staining procedure must not alter the composition of the peptide. Also, and of equal importance, the extraction and staining methods should not introduce contaminants or contribute to background contamination. The extraction procedures should also be rapid and simple in order to allow the simultaneous processing of several peptides. Following SDS-polyacrylamide gel electrophoresis in either one or two dimensional slab or tube formats, proteins and peptides are conventionally stained overnight in 0.2% (w/v) Coomassie® blue in methanol:water:acetic acid (9:9:2). The gel are then destained at 45° in water:methanol:acetic acid (83:10:7).

The stained bands of protein or peptide are carefully cut from the gel using a freshly cleaned blade. Each stained peptide is transferred to a 5 mℓ glass tube containing 2.0 mℓ of 60% formic acid and is homogenized with a Teflon® pestle. The gel particles are sedimented by centrifugation at 2000 xg for 2 min, and the blue, supernatant solution is collected. The particles are washed twice with 1 mℓ aliquots of 60% formic acid, and the supernatant solutions are combined and dried in a stream of nitrogen. One milliliter of 6 *N* HCl is added to the residue and the white precipitate of SDS is removed by centrifugation. The HCl-supernatant solution is transferred to a 5 mℓ conical vial and extracted with 2 mℓ of *n*-octanol. The octanol is added to the vial, mixed thoroughly, centrifuged and the organic phase containing the Coomassie® blue is removed. This procedure is repeated twice. The acid phase, free of SDS and Coomassie® blue, is diluted with 1 mℓ of water and dried under a stream of dry nitrogen.

Control regions of the gels which contain no protein should be recovered and extracted identical to the Coomassie® blue staining regions. Under normal circumstances small amounts of serine, threonine, and alanine are observed in controls. However, the levels of background rarely exceed 5% of any amino acid recovered from peptide-containing sections of the gel.[69] Background levels are linearly related to the volume of the gel controls thus making correction of the analysis simple. When 2 to 100 nmol of protein or peptide are subjected to SDS-electrophoresis followed by extraction in this manner, yields of 85 to 90% are routine.

E. Peptide Alignment

In reconstructing the primary structure of a protein, at least two sets of peptides obtained from cleavage of the protein with different chemical or enzymatic specificity are usually required. A summary of computerized reconstruction of protein structures has been previously presented.[70] Limited chemical and enzymatic cleavages which generate a small number of fragments, often allow reassembly and alignment from direct analyses of each peptide simply based on size, amino acid composition, and amino and carboxyl terminal analyses (Figure 1C).

F. Homology Peptide Mapping

Following the initial separation, pools of peptides from the homologous proteins are subjected to two-dimensional electrophoresis and chromatography on thin-layers of cellulose. Figure 3 shows sets of homologous peptide maps of human and rabbit triosephosphate isomerase tryptic peptides. The peptides are recovered and their amino acid compositions determined by high sensitivity amino acid analysis. The assumption in homology peptide mapping is, if two peptides from the two homologous proteins exhibit identical electrophoretic and chromatographic coordinates and also exhibit identical amino acid compositions, the peptides probably have identical sequences. Thus, it is possible to establish a large portion of the sequence of a homologous protein with little direct sequencing. If single amino acid changes are found in the composition of the homologous peptides, it may be relatively easy to predict the differences in sequence. For example, analysis of a peptide from human triosephosphate isomerase revealed an additional serine and one less threonine when compared to the homologous peptide from rabbit (Figure 4). This case represented a single amino acid substitution, but sequencing was required to determine which of the two threonines had been replaced by serine.

Details of the methods of thin-layer peptide fingerprinting have been presented elsewhere,[71] thus the following is a brief outline of the procedure.

1. Fragmentation Into Small Peptides

The initial steps of sample preparation depend on the state of the material. For example, if the sample is from a polyacrylamide gel, it must be extracted from the gel and concentrated. If the protein contains either sulfhydryls of disulfides, it is necessary to prevent disulfide interchange by blocking these groups by performic acid oxidation,[72] or S-carboxymethylation.[73] Following the required extractions and modifications the peptide or protein is dried in a thickwall, conical glass vial and resuspended in 200 $\mu\ell$ of 0.2 *M* ammonium bicarbonate, pH 8.0. If the protein is not entirely soluble it should be triturated or placed in an ultrasonic chamber to form a finely divided suspension. Trypsin (1% of the total sample) is added at two stages to minimize autodigestion of the protease. The trypsin solution (0.1 mg/mℓ of ice-cold water) is prepared immediately prior to the digestion, and the digestion is allowed to proceed at 37°, while slowly mixing with a magnetic stirrer. After 3 hr a second aliquot of trypsin is added, and the digestion is allowed to proceed for an additional 3 hr. It is important to use trypsin that has been treated to inactivate chymotryptic activity (e.g., L-1-tosylamide-2-phenylethylchloromethyl ketone-treated). After lyophilization, the material is washed from the walls of the vial with 50 $\mu\ell$ of 2% ammonium hydroxide, and the reaction vial centrifuged to concentrate the sample to the bottom of the tube. The solution is dried with a gentle stream of dry nitrogen and redissolved in 10 $\mu\ell$ of 2%

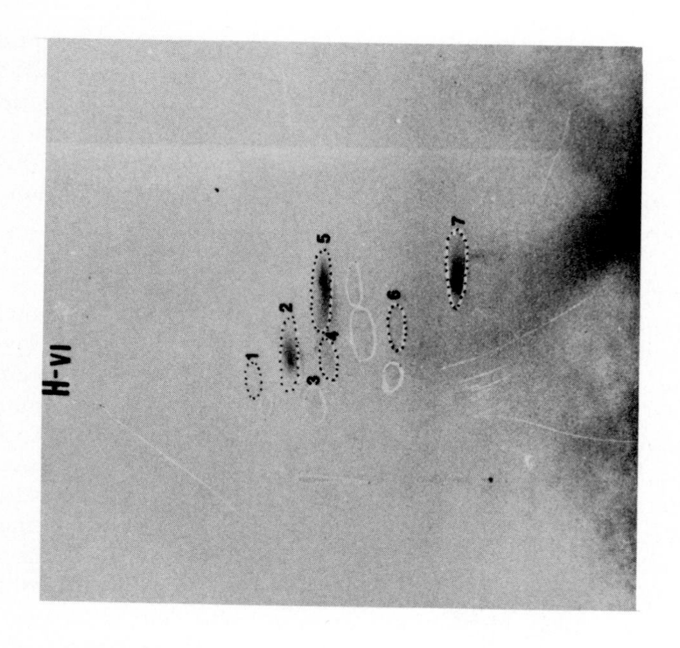
H-VI
1
2
3
4
5
6
7

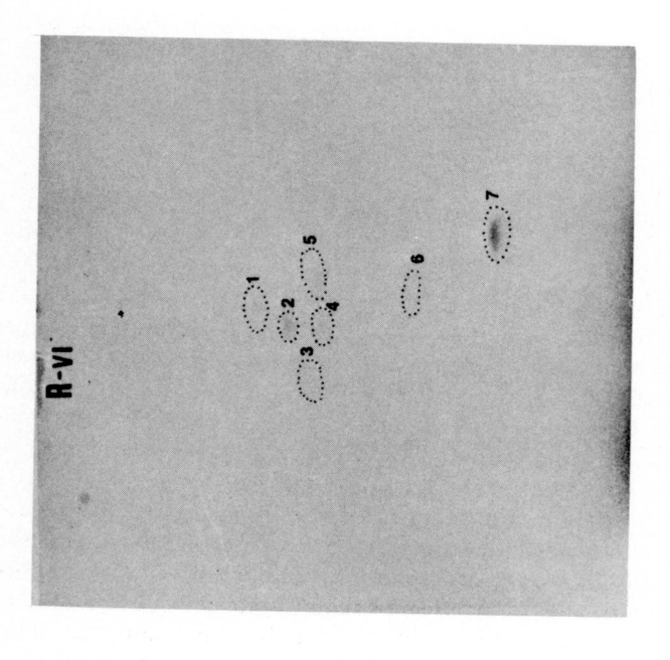
R-VI
1
2
3
4
5
6
7

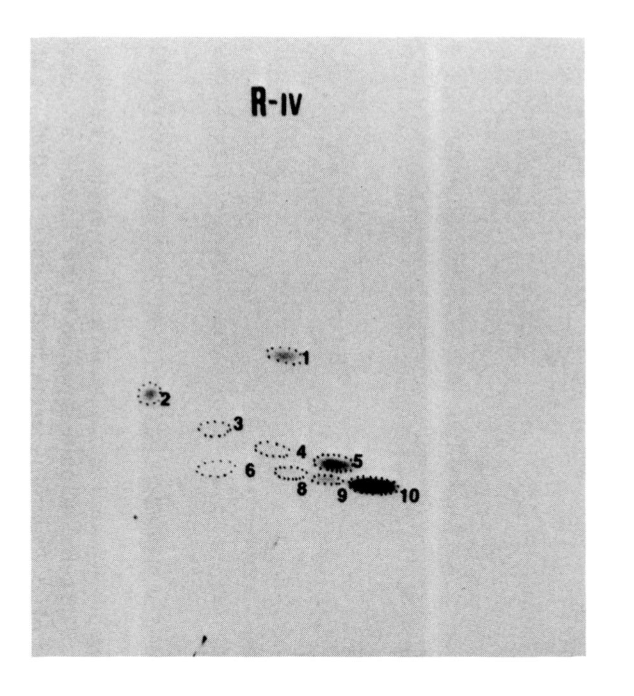

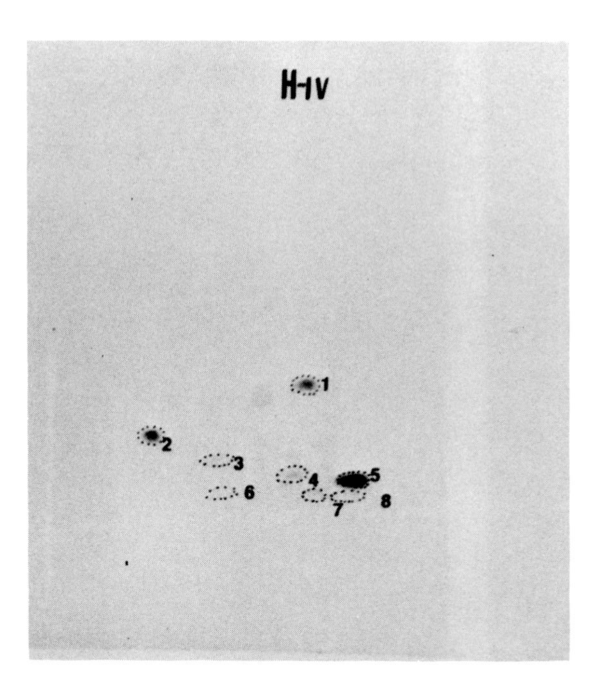

FIGURE 3. Homology peptide mapping of human and rabbit triosephosphate isomerase. Human and rabbit isomerase were isolated and digested with trypsin. Peptides were first separated into seven pools by gel filtration. The top panel shows two-dimensional, thin-layer peptide maps of pool IV (top).

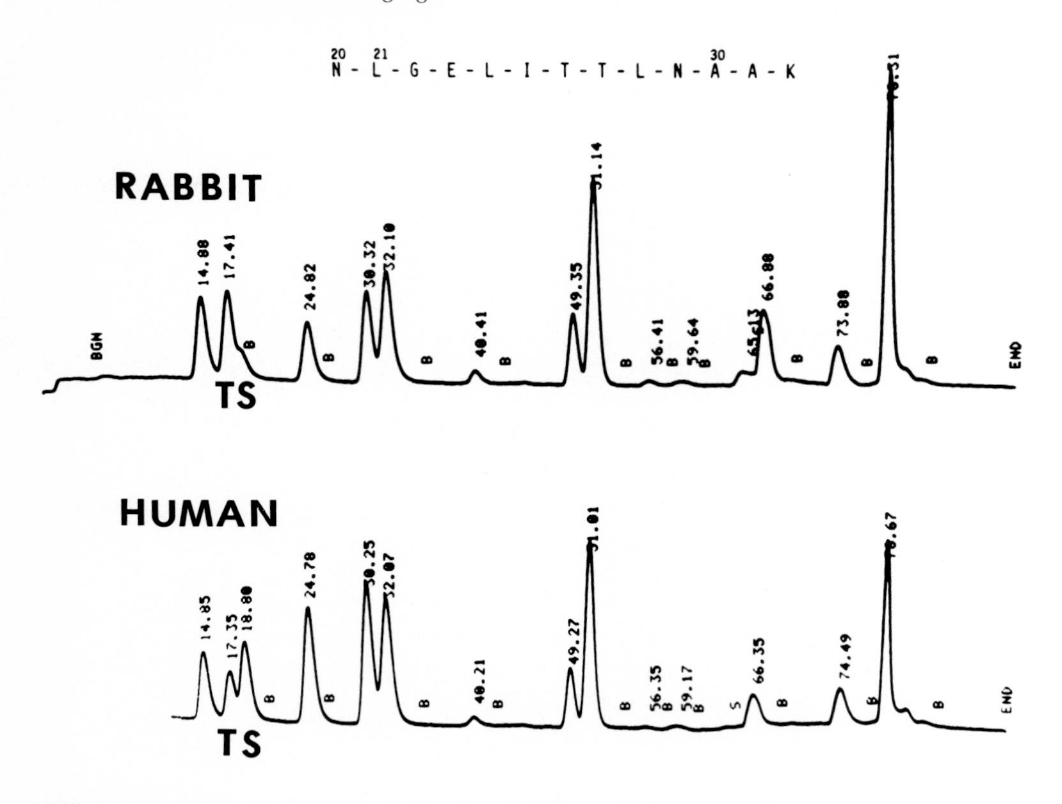

FIGURE 4. Amino acid analysis of homologous peptides recovered from thin-layer peptide maps. Triose-phosphate isomerase from rabbit muscle and human placenta were subjected to homology sequence analysis. After tryptic digestion and gel filtration pool VI was collected from both species and subjected to two-dimensional, thin-layer fingerprinting as shown in Figure 3. The homologous peptides numbered 1 from both species were recovered, hydrolyzed and subjected to high sensitivity amino acid analysis. The only difference is the extra serine (S) and one less threonine (T) in the peptide from the human enzyme. The amino acid sequence of the rabbit peptide is shown above (positions 20-32).

ammonium hydroxide. From 0.2 to 1.0 nmol (1 to 5 $\mu\ell$) of sample is then applied near the corner of the thin-layer sheet 3 cm from either edge. The size of the application spot is important in order to obtain maximal sensitivity and optimal resolution of the peptides, and should not exceed 2 mm in diameter.

2. Electrophoresis

Electrophoresis can be conducted in a varsol cooled immersion system as described by Gracy[71] or with a cooled flatplate system (e.g., Desaga Double Chamber Model). In either case it is essential that the system allow simultaneous electrophoresis of the two homologous samples. The cellulose thin-layer sheets are presprayed with a fine mist of electrophoresis buffer. A small vial (approximately 1 cm in diameter) is held over the application site during the spraying so that the buffer does not directly wet it. The buffer is allowed to slowly wet the application site by capillary action so that the sample is not smeared. The cellulose plate is placed in the chamber and electrophoresis is conducted at either pH 3.7 (pyridine:acetic acid:water = 1:10:89) or at pH 6.5 (pyridine:acetic acid:water = 25:1:225) at constant power (3.5 watts) or constant voltage (300 V, approximately 20 mA). Usually, 2 hr is sufficient for separation of most peptides. After electrophoresis, the thin-layer sheets are removed from the chamber and dried at room temperature for at least an hour in a dust-free area.

3. Chromatography

The thin-layer sheets are turned such that the site of application and the peptides separated by electrophoresis are horizontally distributed across the bottom of the sheet. The thin-layer sheets are placed in a pre-equilibrated chromatography tank, and ascending chromatography is carried out until the solvent (butanol:pyridine:acetic acid:water = 75:50:15:60) front is within 2 cm of the top (3 to 4 hr). The peptide maps are then removed from the chamber and air dried as before.

4. Staining and Visualization of Peptides

The usual choice of reagents for detection of the peptides includes ninhydrin, fluorescamine, and *o*-phthaldialdehyde. *O*-Phthaldialdehyde and ninhydrin permit the detection of free amino acids in the 20 to 500 pmol range, and are superior to fluorescamine.[74] Ninhydrin and fluorescamine, on the other hand, are better for the detection of peptides.[74] Only fluorescamine-treated peptides can be recovered from thin-layers without destruction of α and ε amino groups.

Although some choice of reagents and/or conditions will be predicated by the types of peptides being analyzed, the following general procedure has been successfully employed in our laboratory numerous times for many different types of peptides. The thin-layer plate should first be examined under both long and short wavelength ultraviolet light. This serves two purposes. First, peptides containing a high content of aromatic amino acids or a fluorescent cofactor may be detected prior to reaction with any reagent. Secondly, this permits identification of any fluorescent "artifacts."

The peptide map is then sprayed lightly with a mist of 10% v/v triethylamine in methylene chloride and allowed to dry for a few seconds[75] before spraying with 0.01% fluorescamine (1 mg in 10 mℓ acetone) followed by a second spraying of triethylamine. The maps are then viewed under short wavelength ultraviolet light. In preliminary experiments it is advisable to respray the peptide fingerprint with ninhydrin (0.3% w/v in 30 mℓ of collidine, 100 mℓ of acetic acid, and 870 mℓ of ethanol) and heat with a forced air "heat gun". The purpose of a second spraying with ninhydrin is to locate free amino acids or peptides which are not detected by fluorescamine. However, once the location of peptides has been ascertained, all subsequent studies requiring recovery and analysis should use only fluorescamine detection.

5. Recovery of Peptides from Thin-Layers

After peptides are located on the thin-layers, they are circled with a pencil. The cellulose or silica gel is carefully scraped from the plastic backing and transferred through small funnels to Pasteur pipets (150 mm) which have been tightly plugged with one-fourth of a glass-fiber membrane (20 mm diameter Sartorius SM 13400) and prewashed with 2 mℓ of 6 *N* HCl. Two hundred microliters of 6 *N* HCl containing 0.02% 2-mercaptoethanol is added to each pipet, and after 15 min the HCl is then forced through the filter with nitrogen, and the process is repeated twice. Utilizing this procedure most small peptides can be recovered in yields of 70 to 95%.[74] Recovery is generally better from cellulose than from silica gel, but in both cases background levels of contaminants are quite low and do not interfere with subsequent amino acid analyses at the subnanomole levels. Blank areas should be scraped, extracted, and analyzed as background controls.

G. High Sensitivity Amino Acid Analysis

Several methods have been commonly used for peptide hydrolysis. In the case of small peptides, usually a single hydrolysis time (e.g., 24 hr in 6 *N* HCl at 110°) is sufficient, and only with larger peptides are multiple hydrolysis times required. The incorporation of 2 to 5% (thioglycolic) mercaptoacetic acid or 3-(3-indolyl) propionic acid in 6 *N* HCl or the use of 3 *N p*-toluenesulfonic acid containing 0.2% 3-(2-aminoethyl) indole increase the recoveries of tryptophan.[76-78] Penke et al.[79] reported even better tryptophan yields using 3 *N* mercaptoethanesulfonic acid. The chief drawback to the latter two methods is the need for neutralization with resultant volume changes and formation of precipitates.

Specially designed hydrolysis tubes are available (e.g., Pierce Chemical Co. or Regis Chemical Co.). These tubes are designed to be reusable and feature a sidearm for evacuation and a Teflon® plunger which provides an inert vacuum seal. The simplicity and ease of reusable sidearm tubes makes them very attractive, but problems arise when working at the microgram scale due to background contamination. For microanalyses, specially designed disposable hydrolysis ampules (e.g., Wheaton #651502, 1 mℓ cap.) are preferred. The acid cleaned ampules are stored inverted in a covered acid-cleaned beaker. As the ampules are needed they are removed, and an aliquot of the sample added. The neck of the ampule is covered with parafilm. Small pinholes are punched in the parafilm, the sample is frozen in acetone-dry ice, and lyophilized. After drying, the parafilm is removed and 6 *N* HCl containing 5% thioglycolic acid is added to the ampule. The ampule is frozen in an acetone-dry ice bath and while applying a vacuum, allowed to slowly thaw and degas. After total degassing, the sample is re-frozen and sealed in vacuo. Following hydrolysis at 110° for 24 hr the ampule is snapped open, the sample dried directly in the hydrolysis ampule with a stream of dry nitrogen, and dissolved in 50 to 200 $\mu\ell$ of 0.2 *N* sodium citrate pH 2.2. Analysis of hydrolysates is conducted by standard high sensitivity methods.

H. Amino Terminal Determination

The most common, useful method for amino terminal analysis of proteins and peptides is the dansyl chloride method developed by Hartley,[80] although many other alternatives are available.[81] Aminopeptidase combined with high sensitivity amino acid analysis or high sensitivity degradation procedures (described below) are also very useful. Moreover, in some cases of microsequencing, it is necessary to independently establish the amino terminus. We describe here a general microtechnique for amino terminal determination using dansyl chloride.

A solution of peptide (0.1 to 1 nmol) is transferred into a small (4 × 50 mm) pyrex test tube. After drying in vacuo the peptide is redissolved in 25 $\mu\ell$ of 0.2 *M* $NaHCO_3$, pH 8.5, and the solution is centrifuged and again evaporated to remove traces of ammonia. The residue is redissolved in 25 $\mu\ell$ of deionized, distilled water and an equal volume of dansyl chloride solution (2.5 mg/mℓ in acetone) to give a final concentration of 5 m*M* DNS-chloride, 1 m*M* peptide and 50% acetone. The pH of the solution must be 8.5 to 9.8. Th tube is covered with parafilm, and the reaction is allowed to proceed for 1 hr at 37°, then dried in vacuo. For the labeling of proteins or insoluble peptides, alternative conditions are used: approximately 1 nmol of sample is heated in a boiling water bath in 50 $\mu\ell$ of 1% SDS for 2 to 5 min. After the solution has cooled, 50 $\mu\ell$ of 0.1 *MN*-ethylmorpholine pH 8.5 is added and mixed thoroughly. To the solution is added 100 $\mu\ell$ of dansyl chloride solution (5 mg/mℓ in acetone), and dansylation is allowed to proceed for 1 hr at 37°. Acetone (0.5 mℓ) is added to precipitate the protein, and the tube is inverted several times after covering with parafilm. The protein is pelleted by centrifugation, washed with 0.5 mℓ of 80% acetone and dried.

In either of the above procedures 100 $\mu\ell$ of 6 *N* HCl is added, and the tube is evacuated and sealed. Hydrolysis is carried out at 110° for 16 to 24 hr. The hydrolysate is then dried in vacuo under P_2O_5. The dansyl-amino acids are dissolved in 10 $\mu\ell$ of 50% pyridine for chromatography. The dissolved sample hydrolysate (1 $\mu\ell$) is spotted on a corner of one side of a polyamide sheet which is coated on both sides (Cheng Chin Chem. Co. or Schleicher and Schull Co.). The application spot should not exceed 2 mm in diameter. A standard mixture containing about 0.05 nmol of each dansyl-amino acid is spotted onto the back of the sheet opposite of the unknown. Identification of the dansyl-amino acid is performed by two-dimensional thin-layer chromatography. Ascending chromatography is first carried out with solvent I (1.5% formic acid). The sheet is dried, and then developed in solvent II (benzene:acetic acid; 9:1 v/v) in the direction perpendicular to that with solvent I. Identification of the unknown DNS-anino acid is carried out under ultraviolet illumination by referring the coordinates of the "internal" standards on the opposite side. Identification of DNS-derivatives of aspartate, glutamate, serine and threonine, and histidine should be carried out after a third development with solvent III (ethylacetate-acetic acid-methanol, 20:1:1, v/v) in the same direction as solvent II.

I. Carboxyl Terminal Determination

Kinetic analysis of amino acids liberated by carboxypeptidases A and B[82] or Y[83] are the procedures most widely used in microanalysis of the carboxyl terminal region. A combination of carboxypeptidase A and B can remove all of the protein-constituent amino acids at its carboxyl terminal with the exception of proline. Carboxypeptidase Y has the ability to remove most amino acid residues, including proline. In general, however, when the penultimate and/or terminal residues have aromatic and aliphatic side chains, catalysis is most rapid. The release of terminal histidine, arginine, and lysine is relatively slow.

Commercial preparations of carboxypeptides A or B are normally aqueous suspensions of crystals, which have been treated with diisopropyl fluorophosphate. The suspension is stable at 4° for years. The enzyme is active in crystalline suspension, but usually the crystals are dissolved before use. A suspension containing 1 mg of protein is diluted with 1 mℓ of 0.1 *M* $NaHCO_3$. The crystals are sedimented quickly, and the supernatant solution which may contain free amino acids is discarded. The crystals are dissolved in 0.2 *M* $NaHCO_3$, pH 8.0 (or 0.2 *M* *N*-ethylmorpholine-acetate, pH 8.0). Commercially available carboxypeptidase Y is supplied as a salt-free, lyophilized powder. Almost full activity is restored by storing the solution overnight at 5°. Therefore, the lyophilized enzyme should be dissolved in water or in 10 m*M* sodium phosphate buffer, pH 7.0, 1 day before use.

Prior to digestion, it is necessary to assure denaturation of native proteins and large peptides. Carboxymethylation or performic acid oxidation is preferred when the protein substrates contain disulfide bonds or free sulfhydryls. In addition, the carboxypeptidases are catalytically active in 6 *M* urea or in 0.1% sodium dodecyl sulfate, and thus a denaturant solution may be useful.

Carboxypeptidase A and/or carboxypeptidase B digestions are carried out in 0.2 *M* ammonium bicarbonate (or 0.2 *M* *N*-ethylmorpholine), pH 8.0, while carboxypeptidase Y digestion is performed in 0.1 pyridine-acetate, pH 5.5. For kinetic studies using microscale amino acid analysis the denatured protein or peptide is dissolved to a concentration of approximately 0.1%. The carboxypeptidase is added at an enzyme:substrate ratio of 1/20 to 1/200 (molar basis) and the mixture, in a 1 mℓ conical, screw-capped tube is incubated at 37°. At timed intervals, (e.g., 0, 0.25, 0.5, 1, 2, 4, 8 hr), 0.1 to 5 nmol portions of the mixture are removed and acidified to pH 2 to terminate the reaction. Any precipitate is removed by centrifugation, and the supernatant solution is directly subjected to amino acid analysis. A careful kinetic study can generally elucidate the carboxyl terminal sequence of the unknown sample to at least three penultimate amino acid residues.

Under standard amino acid analysis conditions, asparagine and glutamine overlap serine. For the quantitation of the three amino acids, asparagine and glutamine must be hydrolyzed to aspartate and glutamate and then determined. The hydrogen form of sulfonated polystyrene (Dowex 50-X8, Bio-Rad Laboratories) is added to the reaction mixture until the pH is 2 to 3. The mixture is mixed for 20 min, the supernatant solution removed and the resin beads are washed with two resin volumes of water. The adsorbed amino acids on the resin are extracted three times with 5 *N* NH_2OH and the extracts are evaporated to dryness and hydrolyzed with 5.7 *N* HCl, and subjected to amino acid analysis. The increase in aspartate and glutamate after acid hydrolysis gives an estimation of the amide content.[84]

J. Microsequencing

1. General Methods

A variety of analytical procedures have been developed recently which permit protein and peptide sequencing at the subnanomole levels. Improvements have been achieved by increasing the sensitivity of phenylthiohydantoin detection, by optimizing polypeptide retention in the spinning cup sequencer, and by using solid phase techniques. Application of radiolabeling techniques via in vivo or in vitro radiotracing has also permitted remarkable increases in sensitivity. With the novel designs in automated sequencers and high-performance liquid chromatography (HPLC), Hunkapiller and Hood[85] were able to perform sequencing at the 10-pomol level. Samples derived from HPLC or stained SDS-gels can be thus directly subjected to microsequencing analysis.

The detection of phenylthiohydantoyl amino acids (PTHs) has been traditionally achieved by paper and thin-layer chromatographic methods. Thin-layer chromatography remains an important method of analysis though, it exhibits several drawbacks because the procedure is difficult to quantitate. Gas chromatography has never fully contributed its merits to the detection and quantitation of PTHs. Reconversion of PTHs to their parent amino acids provides an alternative method for detection. Since most laboratories interested in microsequencing will already be equipped with a high sensitivity amino acid analyzer system, this method is particularly attractive. Lai[86] accomplished back hydrolysis of the thiazolinones with acid in the presence of a reducing agent. When his procedure was slightly modified and combined with a simple manual manipulation, microscale sequencing was found to be possible in most cases.[87] The increasing number of developments in HPLC has also made this a most promising technique for increasing the sensitivity in automatic sequence determination.

A number of new reagents in place of Edman's original phenylisothiocyanate (PITC) have also extended the sensitivity of automated or manual sequencing. These compounds include 4-sulfo-phenylisothiocyanate, cyanomethyldithiobenzoate, thioacetylthioglycolic acid, pentafluoro-PITC, 4-*N,N*-dimethylaminoazobenzene 4′-isothiocyanate, and others.[88] Recently, the double coupling technique using *N,N*-dimethylaminoazobenzene isothiocyanate (DABITC) and PITC for manual or automatic sequencing has been developed.[89,90] The method takes advantage of the intensely colored DABTH derivatives which show a high absorption in the visible region (340 to 580 nm), such that 5 to 25 pmol of the derivatives can be easily detected.

Microtechniques for sequence determination, using the spinning cup sequencer have been discussed previously.[91] Its importance in protein sequencing is obvious, however, the cost of the instrument and the expense for maintenance of the liquid phase sequencer may preclude it from some laboratories.

The use of solid phase methods of protein and peptide sequencing pioneered by Laursen[92] and colleagues, has been reviewed in similar detail.[93-98] The solid phase sequencer was originally intended for peptides containing 30 amino acid residues or less, but the improved technology of amino-glass supports has made possible the solid phase sequencing of proteins as well as peptides of any size.[98,99] Since solid phase sequencing depends on having the peptide covalently linked to an insoluble resin support and then pumping the reagents over the immobilized peptide, the linking of the peptide to the support is a critical step. Unfortunately, there is no single procedure which will work in all cases. However, using one of the several methods, most peptides, and even proteins, can be attached and sequenced.[98,99] The two most useful resins for peptide sequencing are aminopolystyrene, for diisothiocyanate coupling of peptides, and triethylenetetramine (TETA) resin for all other procedures. For the attachment of proteins, isothiocyanate glass is recommended. For aminopolystyrene there seems to be an upper limit of 30 to 40 residues, for the TETA resin about 70 residues, for the attachment of peptides. The coupling procedures involve linking an amine group of the resin to the terminal carboxyl of the peptide. This frequently requires the blocking of other groups, or, when carboxyls are not available, the modification of selected groups to make them appropriately reactive. Carboxyl-terminal lysine or amino ethylcysteine can be coupled to aminopolystyrene resin through their amino side chains by treatment with *p*-phenylenediisothiocyanate. Arginine terminal peptides can be converted to ornithine with 50% hydrazine and then coupled as for lysine. Peptides generated by cyanogen bromide cleavage of proteins can be sequenced by conversion of the C-terminal homoserine to the lactone and coupling it to TETA resin. Carboxyl groups can be linked to TETA resin by the use of water soluble carbodiimides, especially *N*-dimethylaminopropyl-*N*-ethycarbodiimide. After the coupling has been completed, the resin is packed into a column for automated sequencing. The solid phase sequencer is much less expensive to operate, requiring far less reagents than the liquid phase instrument, and it is also well suited for microsequencing. A separate pump is preferred to deliver radioactive PITC (either ^{35}S or ^{3}H), and then the reaction is driven to completion with excess of nonradioactive PITC. The radioactive phenylthiohydantoin is collected in tubes which contain a mixture of all 20 nonradioactive phenylthiohydantoins. The identification of the radioactive PTH is easily achieved by thin-layer chromatography with the nonradioactive derivatives serving as internal markers. The radioactive PTH can be located either by radioautography or by scraping the spots from the plate and analyzing in a liquid scintillation counter.

We describe here the use of simple manual microsequencing using either DABITC as a coupling agent or back hydrolysis as the detection method.

2. Manual Sequencing Methods

a. Microsequencing Procedure Using Back Hydrolysis

All reagents should be the equivalent of "sequanal grade" as provided by Pierce Chemical Company. Pyridine should be distilled at least twice over KOH and ninhydrin. The 50% PITC solution is prepared every 2 weeks and stored frozen under nitrogen. From 1 to 10 nmol of peptide are dissolved in 120 $\mu\ell$ of 75% aqueous pyridine. Ten microliters of 50% phenylisothiocyanate (PITC, dissolved in 75% pyridine, pH 9.0) are added to a 1-mℓ thickwalled conical vial. The vial is flushed with dry nitrogen, sealed and incubated at 52° for 60 min. Excess PITC is extracted with 1-chlorobutane or benzene-ethylacetate (2:1), and the derivatized peptide in vacuo. Cleavage is achieved by adding 100 $\mu\ell$ of anhydrous trifluoroacetic acid and incubating for 15 min at 52°. After drying, the thiazolinone derivative is extracted with chlorobutane and used for back hydrolysis and analysis. The aqueous phase is dried and submitted to the next cycle. The thiazolinone derivatives are dried in vacuo and hydrolyzed in 200 $\mu\ell$ of 6 *N* HCl containing either 1% 2-mercaptoethanol and 0.2% phenol or 0.1% $SnCl_2$. Details of the recovery of amino acids from thiazolinones have been compared previously.[87] The method using 6 *N* HCl containing mercaptoethanol and phenol is superior in most cases. However, if serine and threonine are suspected, the $SnCl_2$/HCl method is advisable since they can be reconverted to alanine and α-aminobutyric acid, respectively.

This method has the advantage of simplicity, requiring no specialized equipment other than an update amino acid analyzer and is excellent for peptides up to 15 residues. Another advantage is the sensitivity and accurate quantitation at which each sequencing step can be conducted. The method has the inherent disadvantage of not distinguishing glutamate and asparatate from their respective amides.

b. Microsequencing Using N,N-Dimethylaminoazobenzene 4′-Isothiocyanate (DABITC)

DABITC can be obtained from Pierce Chemical Company. Glacial acetic acid saturated with HCl is prepared by bubbling HCl gas through acetic acid 2 hr.

Sequencing analysis using DABITC double coupling was primarily developed by Chang et al.[89] We have adapted their original protocol with minor modifications for our routine microsequencing. Peptides or proteins (2 to 8 nmol) are dried in an acid-washed conical 1-mℓ vial fitted with a Teflon® stopper. The sample is then dissolved in 80 μℓ of aqueous 50% pyridine-H_2O and 48 μℓ of DABITC (10 nmol/μℓ, freshly prepared) is added. The vial is flushed with dry nitrogen for 10 sec, sealed and placed in a heating block at 55° for 50 min. After the first coupling, 10 μℓ PITC is added and the second coupling allowed to proceed at 55° for 30 min. The excess reagents and byproducts are removed by mixing the solution with 0.5 mℓ heptane/ethyl acetate (2:1, v/v) on a vortex mixer and centrifuging. The organic phase is removed and discarded. This process is repeated twice more. The reaction mixture is dried in vacuo with precautions to prevent bumping. The derivatized peptide is then subjected to cleavage by dissolving in 100 μℓ anhydrous trifluoroacetic acid, flushing with N_2, sealing and heating at 60° for 15 min. The sample is then dried in a vacuum desiccator and dissolved in 50 μℓ of water. Extraction of the thiazolinones DABTZ-amino acids and PTZ-amino acids) is achieved by mixing with 200 μℓ of butyl acetate on a vortex mixer and centrifuging. After removal of the butyl acetate extract, the peptide in the aqueous phase is dried in a desiccator and subjected to the next degradation cycle. The butylacetate extract is evaporated and redissolved in water (20 μℓ) and acetic acid saturated with HCl (40 μℓ). Conversion of the thiazolinones into thiohydantoins is carried out by heating at 52° for 50 min. The sample is dried and redissolved in 5 to 30 μℓ of ethanol. The identification of DABTH-amino acids is carried out on polyamide sheets (5 × 5 cm or 2.5 × 2.5 cm). The blue synthetic marker, DABTC-diethylamine (10 to 20 pmol), is co-chromatogrammed with each unknown to facilitate identification. The first dimension is developed in acetic acid-water (1:2, v/v), and toluene-n-hexane-acetic acid (2:1:1; v/v) is used as solvent for second dimensional separation. Only DABTH-Leu and DABTH-Ile are unseparable under these conditions. For identification of these two residues, chromatography on silica gel plates can be performed with chloroform-ethanol (100:3, v/v) as developing solvent, or back hydrolysis can unambiguously identify the residue. After separation the plates are dried and exposed to HCl vapor. The sensitivity is 10 to 50 pmol.

The advantages of this microsequencing procedure are a single polyamide TLC identification is adequate to resolve all the common amino acids as their colored DABTH derivatives except for leucine and isoleucine, and the color difference between the amino acid-DABTH derivatives (red) and contaminants (blue) greatly facilitates the identification.

3. Detection of Phenylthiohydantoin Amino Acids by High Performance Liquid Chromatography

None of the methods using gas-liquid chromatography or thin-layer chromatography for separation of PTHs is completely satisfactory. In spite of the fact that additional specialized instrumentation is required, HPLC has many meritorious features:

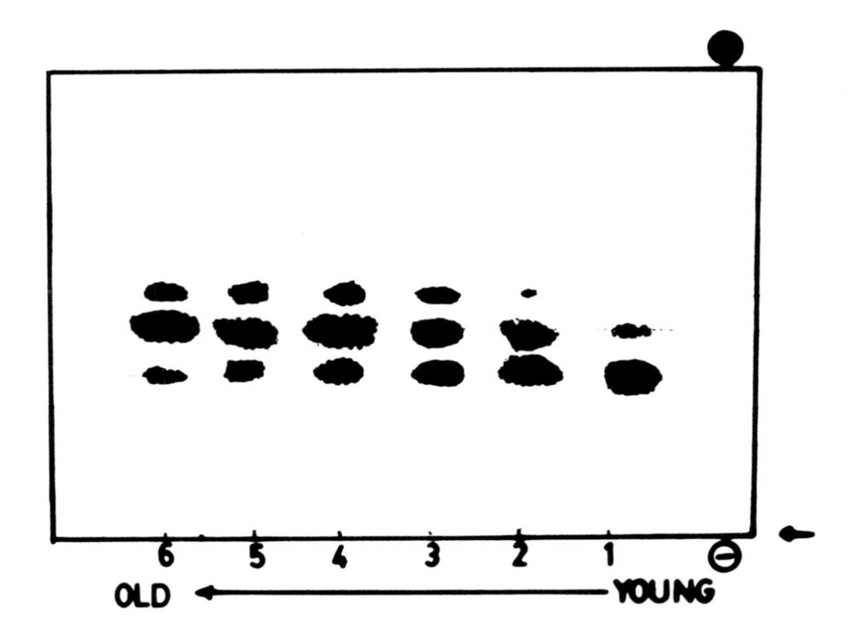

FIGURE 5. Electrophoresis of human triosephosphate isomerase from erythrocytes of different age. Peripheral blood was drawn by standard venipuncture and fractionated into six fractions according to age by density gradient centrifugation. The reticulocytes were applied to the right with the older cells to the left. Electrophoresis was toward the anode. The gel was stained with an enzyme activity stain specific for TPI. The figure is from a direct tracing of the gel.

all the PTHs are stable during analysis and have similar absorbancy in the ultraviolet, the sensitivity for detection is at the picomole level. All the common PTHs are eluted from a single column, and less than 30 min is required for an analysis. Methods have been developed by many research groups.[100-103] The systems of Bhown et al.[102] and Johnson et al.[103] using reverse phase Bondapak C_{18} (Waters Associates) and Zorbax CN columns (Dupont) are being used in this laboratory.

IV. MOLECULAR BASIS FOR ABNORMAL TRIOSEPHOSPHATE ISOMERASE DURING AGING

Most human tissues contain a primary form of TPI with an isoelectric point of 5.8 designated at TPI-B. However, more acidic forms (TPI-A) are observed in most tissues.[104,105] When human erythrocytes are fractionated according to age a dramatic shift in the relative content of the TPI isozymes is observed. Figure 5 shows that while TPI-B predominates in reticulocytes and young erythrocytes, it is replaced by the more acidic isozymes in "old" erythrocytes. Similarly, Skala-Rubinson et al.[106] reported the accumulation of the more acidic forms of TPI during aging of the eye lens. Recently Yuan et al.[35] demonstrated that the acidic forms of human TPI could be generated in vitro from pure TPI-B. These in vitro forms were shown to be identical to those observed in vivo.

Human TPI was first digested with trypsin and subfractionated by gel filtration into seven peptide pools. Each of these pools of peptides was subjected to two-dimensional homology peptide mapping with rabbit TPI as shown in Figure 3. Two pairs of peptides, H-IV-3/H-IV-6 and H-VI-2/H-VI-4, were found in which each respective pair exhibited identical amino acid compositions. Further structural analysis showed that peptides 3 and 6 from Pool IV represented a tetrapeptide (position 14-17) with the only difference being that peptide 6 contained an asparagine at position 15 rather than an aspartic acid for peptide 3. Similarly, in Pool VI peptides 2 and 4 were identical representing sequence position (69-85) with the single deamidation of asparagine 71 accounting for peptide 4. The two deamidated peptides were observed in peptide maps of the "aged" enzyme formed both in vivo and in vitro. From structural studies such as those described above it was found that the only residues in the enzyme which readily undergo deamidation either in vivo or in vitro are Asn-15 and Asn-71. Additional studies suggested that only after Asn-71 deamidates does Asn-15 subsequently deamidate. Figure 6 shows a three-dimensional model for human TPI constructed utilizing homology analysis and high resolution X-ray information from the chicken enzyme.[107] The model is shown with only the overall secondary and tertiary structure and without the side groups. As seen from the model, Asn-71 from one subunit is juxtaposed to Asn-15 of the other subunit. The deamidation of Asn-15 and Asn-71 thus introduces four new negative charges into the subunit-subunit contact sites. The introduction of two sets of charge-charge repulsions into the subunit contact sites might be expected to facilitate subunit-subunit dissociation. In vitro denaturation titration studies[35] confirmed that, indeed, deamidation of these two specific residues renders the enzyme much more easily dissociable.

Thus, it appears that the specific deamidations of Asn-71 and Asn-15 occur both in vivo and in vitro and, although having little effect on the catalytic activity the enzyme, destabilize the subunit-subunit interactions. In this way a spontaneous dissociation of the enzyme is favored. The protease susceptibility of the dissociated enzyme is much greater than for the native enzyme. A sequential model is shown in Figure 7 whereby the spontaneous deamidation of Asn-71 followed by Asn-15 gives rise to the more acidic subforms of TPI which lead to dissociation and proteolytic digestion.

Although most tissues contain a small amount of the deamidated forms, they appear to be present only in low steady-state levels. On the other hand, in several "aging model systems" the deamidated forms accumulate. Why do the acidic subforms of TPI accumulate during aging? In this case it does not appear to be due to any of the originally proposed mechanisms (i.e., infidelity in protein synthesis or abnormal post-synthetic modification). Indeed, the protein is synthesized normally and undergoes its "normal" deamidation. A defect in the activity or specificity of proteases to hydrolyze the deamidated TPI could account for the accumulation of the deamidated forms during aging. Obviously, such a defect at the level of the proteases could also account for the observed accumulation of many other abnormal proteins during aging.

ACKNOWLEDGMENT

This work was supported in part by research grants from the National Institutes of Health (AM14638) (AGO1274) and the Robert A. Welch Foundation (B-502).

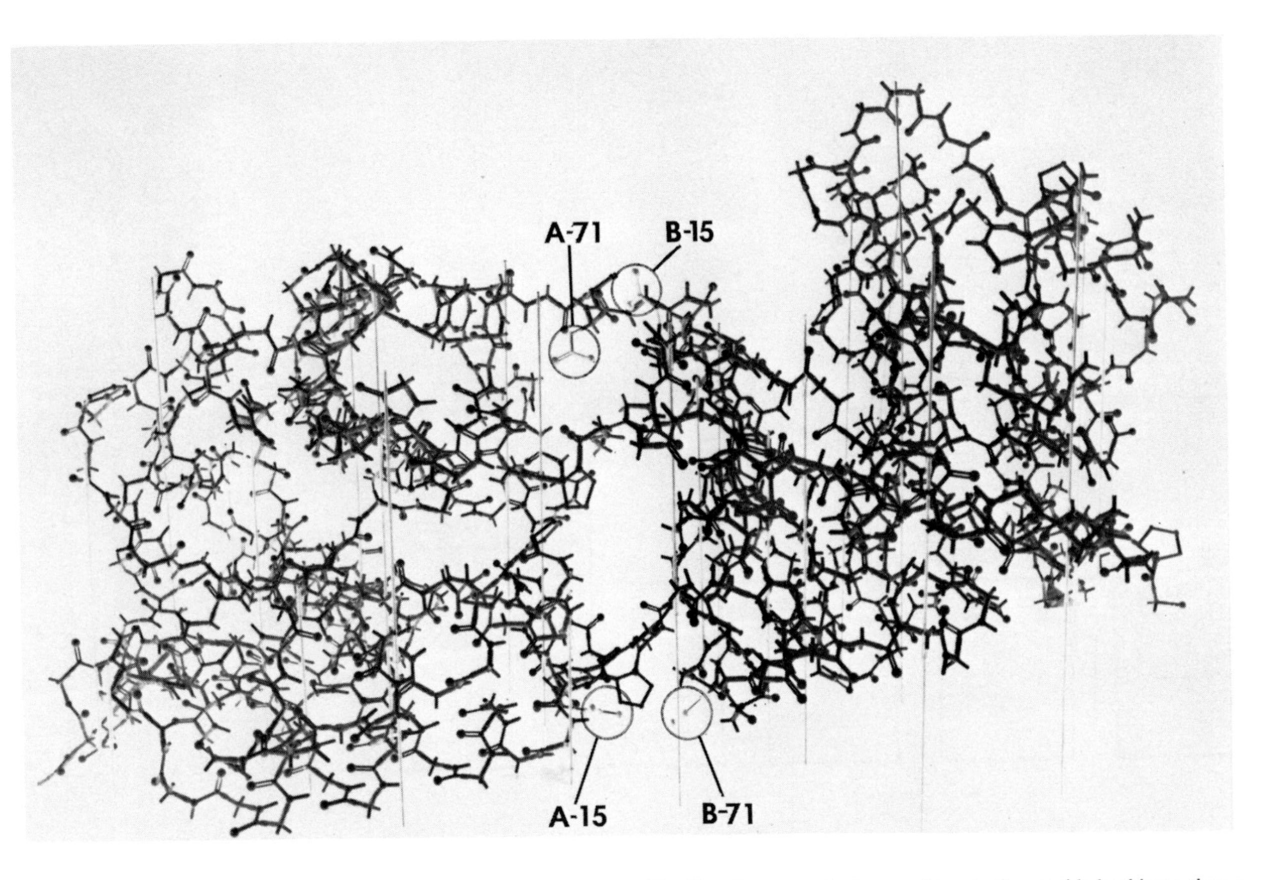

FIGURE 6. Positions of deamidation of triosephosphate isomerase. The dimeric enzyme is shown with only the peptide backbone shown. The subunit on the left has been arbitrarily designated as subunit-A and the subunit on the right designated as B. The contact site between the two subunits can clearly be seen. The positions of spontaneous deamidation of asparagine are shown on the ''A'' subunit as A-15 (Asn-15) and A-71 (Asn-71) and on the ''B'' subunit as B-15 (Asn-15) and B-71 (Asn-71).

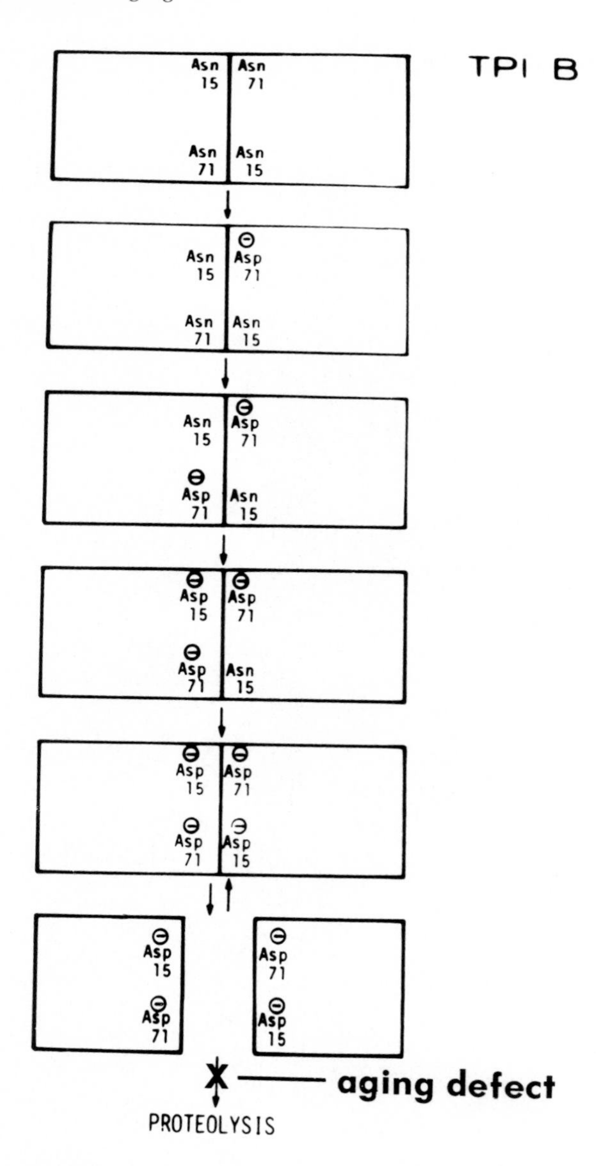

FIGURE 7. Explanation of the accumulation of "abnormal" forms of TPI during aging. The two subunits of TPI are depicted with the juxtaposed asparagines which spontaneously deamidate. The mechanism shows first the deamidation of Asn-71 followed by deamidation of Asn-15 on each subunit. The deamidated forms account for the four acid sub-forms of active TPI observed in many tissues. The deamidations cause the enzyme to dissociate into monomers which are much more susceptible to proteolytic catabolism. In aging cells a defect in the proteolytic system causes the normal intermediates of protein catabolism (i.e., the deamidated forms) to accumulate.

REFERENCES

1. Holliday, R. and Tarrant, G. M., Altered enzymes in ageing human fibroblast, *Nature (London)*, 238, 26, 1972.
2. Gershon, H. and Gershon, D., Detection of inactive enzyme molecules in ageing organisms, *Nature (London)*, 225, 1214, 1970.
3. Gershon, H. and Gershon, D., Inactive enzyme molecules in ageing mice: liver aldolase, *Proc. Natl. Acad. Sci. USA*, 70, 909, 1973.
4. Gershon, H. and Gershon, D., Altered enzyme molecules in senescent organisms: mouse muscle aldolase, *Mech. Ageing Dev.*, 2, 33, 1973.
5. Zeelon, P., Gershon, H., and Gershon, D., Inactive enzyme molecules in ageing organisms nematode fructose-1,6-diphosphate, *Biochemistry*, 12, 1743, 1973.
6. Rothstein, M., Ageing and the alteration of enzymes: a review, *Mech. Ageing Dev.*, 4, 325, 1975.
7. Hayflick, L., The cell biology of human ageing, *Sci. Am.*, 242, 58, 1980.
8. Wulf, J. H. and Cutler, R. G., Altered protein hypothesis of mammalian ageing processed-I, thermal stability of glucose-6-phosphate dehydrogenase in C57BL/6J mouse tissue, *Exp. Gerontol.*, 10, 101, 1975.
9. Reznich, A. Z. and Gershon, D., Age related alterations in purified fructose-1,6-diphosphate aldolase form the nematode turbatrix aceti, *Mech. Ageing Dev.*, 6, 345, 1977.
10. Beauchene, R. E., Roeder, L. M., and Barrows, C. H., The relationships of age, tissue protein synthesis, and proteinuria, *J. Gerontol.*, 25, 359, 1970.
11. Buetow, D. E. and Gandhi, P. S., Decreased protein synthesis by microsomes isolated from senescent rat liver, *Exp. Gerontol.*, 8, 243, 1973.
12. Gritton, G. W. and Sherman, F. G., Altered regulation of protein synthesis during ageing as determined by in vitro ribosomal assays, *Exp. Gerontol.*, 10, 67, 1975.
13. Geary, D. and Florini, J. R., Effects of age of protein synthesis in isolated perfused mouse hearts, *J. Gerontol.*, 27, 325, 1972.
14. Ove, P., Obenrader, M., and Lansing, A., Synthesis and degradation of liver proteins in young and old rats, *Biochim. Biophys. Acta*, 277, 211, 1972.
15. Van Bezooijen, C. F. A., Grell, T., and Knook, D. L., The effect of age on protein synthesis by isolated liver parenchymal cells, *Mech. Ageing Dev.*, 6, 293, 1977.
16. Bradley, M. O., Dice, J. F., Hayflick, L., and Schimke, R. T., Fiber autoradiography of replicating yeast DNA, *Exp. Cell Res.*, 96, 103, 1975.
17. Bradley, M. O., Hayflick, L., and Schimke, R. T., Protein degradation in human fibroblasts (WI-38), *J. Biol. Chem.*, 251, 3521, 1976.
18. Finch, C. E., Enzyme activities, gene function and ageing in mammals, (review), *Exp. Gerontol.*, 7, 52, 1972.
19. Dreyfus, J. C., Kahn, A., and Schapira, F., Posttranslational modifications of enzymes, *Curr. Top. Cell. Regul.*, 14, 243, 1978.
20. Orgel, L. E. The maintenance of the accuracy of protein synthesis and its relevance to ageing, *Proc. Natl. Acad. Sci. USA*, 49, 517, 1963.
21. Mennecier, F. and Dreyfus, J. C., Molecular ageing of fructose-biophosphate aldolase in tissues of rabbit and man, *Biochim. Biophys. Acta*, 364, 320, 1974.
22. Gershon, D. and Gershon, H., An evaluation of the 'error catastrophe' theory of ageing in the light of recent experimental results, *Gerontology*, 22, 212, 1976.
23. Dreyfus, J. C., Rubinson, H., Schapira, F., Weber, A., Marie, J., and Kann, A., Possible molecular mechanisms of ageing, *Gerontology*, 23, 211, 1977.
24. Elens, A. and Wattiaux, R., Age-correlated changes in lysosomal enzyme activities: an index of ageing? *Exp. Gerontol.*, 4, 131, 1969.
25. Robbins, E., Levine, E. M., and Eagle, H., Morphologic changes accompanying senescences of cultured human diploid cells, *J. Exp. Med.*, 131, 1211, 1970.
26. Comolli, R., Hydrolase activity and intracellular pH in liver, heart and diaphragm of ageing rats, *Exp. Gerontol.*, 6, 219, 1971.
27. Lipetz, J. and Cristofalo, V. J., Ultrastructural changes accompanying the ageing of human diploid cells in culture, *J. Ultrastruct. Res.*, 39, 43, 1972.
28. Brock, M. A. and Hay, R. J., Comparative ultrastructure of which fibroblasts in vitro at early and late stages during their growth span, *J. Ultrastruct. Res.*, 36, 291, 1971.
29. Brucnk, U., Ericsson, J. L., Ponten, J., and Westermark, B., On vitro differentiation and specialization of cell surfaces in human glia-like cells, *Acta Pathol, Microbiol., Scand.*, 79, 309, 1971.
30. Bridgen, J., High sensitivity sequence determination of immobilized peptides and proteins, *Meth. Enzymol.*, 47, 321, 1977.

31. Bridgen, J., High sensitivity amino acid sequence determination, *Biochemistry,* 15, 3600, 1976.
32. Bridgen, P. J., Cross, G. A. M., and Bridgen, J., N-terminal amino acid sequences of variant-specific surface antigens from trypanosoma brucei, *Nature (London),* 263, 613, 1976.
33. Yuan, P. M., Talent, J. M., and Gracy, R. W., Elucidating protein structures at the subnanomole level, *Texas J. Sci.,* 31, 43, 1979.
34. Yuan, P. M., Talent, J. M., and Gracy, R. W., Elucidation of the sequence of human triosephosphate isomerase by homology peptide mapping: a rapid method for structural analysis of human enzymes at the nanomole and subnanomole levels, *Biochim. Biophys. Acta,* 671, 211, 1981.
35. Yuan, P. M., Talent, J. M., and Gracy, R. W., Spontaneous deamidation of two specific asparagines in the subunit-subunit contact sites of triosephosphate isomerase: a mechanism for triggering the catabolism of the enzyme, *Mech. Ageing Dev.,* 17, 151, 1981.
36. Tilley, B. E., Izaddoost, M., Talent, J. M., and Gracy, R. W., Recovery and analysis of peptides from thin layer cellulose, *Anal. Biochem.,* 62, 281, 1974.
37. Petell, J. K., Lebherz, H. G., Properties and metabolism of fructose diphosphate aldolase in livers of "old" and "young" mice, *J. Biol. Chem.,* 254, 8179, 1979.
38. Tollefsbol, T. and Gracy, R. W., Proteolytic modifications of human phosphoglycerate kinase from lymphoblasts, *Arch. Biochem. Biophys.,* 205, 280, 1980.
39. Lacko, A. G., Brox, L. W., Gracy, R. W., and Horecker, B. L., The carboxyterminal structure of rabbit liver aldolase (aldolase B), *J. Biol. Chem.,* 245, 2140, 1970.
40. Ramachandran, L. K. and Witkop, B., N-Bromosuccinamide cleavage of peptides, *Meth. Enzymol.,* 11, 283, 1967.
41. Konigsberg, W. H. and Steimamen, H. M., Strategy and methods of sequence analysis, *The Proteins,* 3, 1, 1977.
42. Walsh, K. A., Ericsson, L. H., and Tatani, K., Strategies of amino acid sequence analysis, in *The Versatility of Proteins,* Li, C. H., Ed., Oxford Press, New York, 1978.
41. Konigsberg, W. H. and Steimamen, H. M., Strategy and methods of sequence analysis, *The Proteins,* 3, 1, 1977.
42. Walsh, K. A., Ericsson, L. H., and Tatani, K., Strategies of amino acid sequence analysis, in *The Int. Symp. of Proteins,* Li, C. H. and Neurath, H., Eds., Oxford Press, New York, 1978.
43. Spande, T. F., Witkop, B., Degani, Y., and Patchornik, A., Selective cleavage and modification of peptides and proteins, *Adv. Protein Chem.,* 24, 97, 1970.
44. Butler, P. J. G. and Hartley, B. S., Maleylation of amino groups, *Meth. Enzymol.,* 25, 191, 1972.
45. Omenn, G. S., Fontana, A., and Anfinsen, C. B., Modification of the single tryptophan residue of staphylococcal nuclease by a new mild oxidizing agent, *J. Biol. Chem.,* 245, 1895, 1970.
46. Ozols, J. and Gerard, C., Cleavage of trytophan bonds in cytochrome B_5 by CNBr, *J. Biol. Chem.,* 252, 5986, 1977.
47. Mahoney, W. C. and Hermodson, M. A., High yield cleavage of tryptophanyl bonds by *o*-iodosobenzoic acid, *Biochemistry,* 18, 3810, 1979.
48. Savige, W. E. and Fontana, A., Cleavage of the tryptophanyl peptide bond by dimethyl sulfoxide hydrobromic acid, *Meth. Enzymol.,* 47, 459, 1977.
49. Landon, M., Cleavage at asp-pro bonds, *Meth. Enzymol.,* 47, 145, 1977.
50. Bornstein, P. and Balian, G., Cleavage at Asn-Gly bonds with hydroxylamine, *Meth. Enzymol.,* 47, 132, 1977.
51. Titani, K., Koide, A., Hermann, J., Ericsson, L. H., Santosh, K., Wade, R. D., Walsh, K. A., Neurath, H., and Fisher, E. H., Complete amino acid sequence of rabbit muscle glycogen phosphorylase, *Proc. Natl. Acad. Sci. USA,* 74, 4762, 1977.
52. Lu, H. S., Talent, J. M., and Gracy, R. W., Chemical modification of critical catalytic residues of lysine, arginine and tryptophan in human glucosephosphate isomerase, *J. Biol. Chem.,* 256, 785, 1981.
53. Otieno, S., Generation of a free α-amino group by Raney Nickel after 2-nitro-5-thiocyanobenzoic acid cleavage at cysteine residues: application to automated sequencing, *Biochemistry,* 17, 5468, 1978.
54. Jacobson, G. R., Schaffer, M. H., Stark, G. R., and Vanaman, Specific cleavage in high yield at the amino peptide bonds of cysteine and cystine residues, *J. Biol. Chem.,* 248, 6583, 1973.
55. Degani, Y. and Patchornik, A., Cyanylation of sulfhydryl groups by 2-nitro-5-thiocyanobenzoic acid, *Biochemistry,* 13, 1, 1974.
56. Mitchell, W. M., Cleavage at arginine residues by clostripain, *Meth. Enzymol.,* 47, 165, 1977.
57. Drapeau, G. R., Cleavage at glutamic acid with staphylococcal protease, *Meth. Enzymol.,* 47, 189, 1977.
58. Smith, E. L., Reversible blocking at arginine by cyclohexanedione, *Meth. Enzymol.,* 47, 156, 1977.
59. Porter, R. R., The hydrolysis of rabbit y-globulin and antibodies with crystalline papain, *J. Biochem.,* 73, 119, 1959.

60. **Edelman, G. M., Cunningham, B. A., Gall, W. E., Gottlieb, P. D., Rutishauser, U., and Waxdal, M. J.,** The covalent structure of an entire gamma G immunoglobulin molecule, *Proc. Natl. Acad. Sci. USA,* 63, 78, 1969.
61. **Richards, R. M. and Vithayathia, P. J.,** The preparation of subtilisinmodified ribonuclease and the separation of the peptide and protein components, *J. Biol. Chem.,* 234, 1459, 1959.
62. **Raibaud, O. and Goldberg, M. E.,** Characterization of two complementary polypeptide chains obtained by proteolysis of rabbit muscle phosphorylase, *Biochemistry,* 12, 5154, 1973.
63. Linderström-Lang, K., The initial phases of the enzymatic degradation of proteins, *Proteins and Enzymes,* 6, 100, 1952.
64. **Maizel, J. V., Jr.,** *Methods in Virology,* Vol. 5, Maramorosch, K. and Koprowski, H., Eds., Academic Press, New York, 1971, 179.
65. **Weber, K. and Osborn, M.,** *The Proteins,* 3rd ed., Neurath, H. and Hill, R. L., Eds., Academic Press, New York, 1975, 179.
66. **Fowler, A. V. and Zobin, I.,** The amino acid sequence of beta-galactosidase of *Escherichia coli, Proc. Natl. Acad. Sci. USA,* 74, 1507, 1977.
67. **Rubinstein, M.,** Preparative high performance liquid partition chromatography of proteins, *Anal. Biochem.,* 98, 1, 1979.
68. **Ui, N.,** Rapid estimation of molecular weights of protein polypeptide chain using high-performance liquid chromatography in 6M guanidine hydrochloride, *Anal. Biochem.,* 97, 65, 1979.
69. **Gibson, D. R. and Gracy, R. W.,** Extraction of proteins and peptides from coomassive blue-stained sodium dodecyl sulfate gels, *Anal. Biochem.,* 96, 352, 1979.
70. **Goldstone, A. D. and Needleman, S. B.,** *Protein Sequence Determination,* Springer-Verlag, New York, 1975, 298.
71. **Gracy, R. W.,** Two dimensional thin-layer methods, *Meth. Enzymol.,* 47, 195, 1977.
72. **Hirs, C. H. W.,** Performic acid oxidation, *Meth. Enzymol.,* 11, 197, 1967.
73. **Crestfield, A. M., Moore, S., and Stein, W. H.,** The preparation and enzymatic hydrolysis of reduced and S-carboxymethylated proteins, *J. Biol. Chem.,* 238, 622, 1963.
74. **Schiltz, E., Schnackerz, K. D., and Gracy, R. W.,** Comparison of ninhydrin, fluorescamine and *o*-phthaldialdehyde for the detection of amino acids and peptides and their effects on the recovery and composition of peptides from thin-layer fingerprints, *Anal. Biochem.,* 79, 33, 1977.
75. **Felix, A. M. and Jimenez, M. H.,** Usage of fluorescamine as a spray reagent for the thin-layer chromatography, *J. Chromatogr.,* 89, 361, 1974.
76. **Matsubara, H. and Sasaki, R. M.,** High recovery of tryptophan from acid hydrolysate of proteins, *Biochem. Biophys. Res. Commun.,* 35, 175, 1969.
77. **Liu, T-Y. and Chang, Y. H.,** Hydrolysis of proteins with *p*-toluenesulfonic acid, *J. Biol. Chem.,* 246, 2842, 1971.
78. **Gruen, L. C. and Nicholls, P. W.,** Improved recovery of tryptophan following acid hydrolysis of proteins, *Anal. Biochem.,* 47, 348, 1972.
79. **Penke, B. Ferenzi, R., and Kovacs, K.,** A new acid hydrolysis method for determining tryptophan in peptide and proteins, *Anal. Biochem.,* 60, 45, 1974.
80. **Hartley, B. S.,** Strategy and tactics in protein chemistry (the first BDH lecture), *Biochem. J.,* 119, 805, 1970.
81. **Narita, K., Matsuo, H., and Nakajima, T.,** *Protein Sequence Determination,* Springer-Verlag, New York, 1975, 30.
82. **Ambler, R. P.,** Enzyme hydrolysis with carboxypeptidases, *Meth. Enzymol.,* 25, 143, 1972.
83. **Hayashi, R.,** Carboxypeptidase Y in sequence determination of peptides, *Meth. Enzymol.,* 47, 84, 1977.
84. **Thompson, A. R.,** C-terminal residue of lysozyme, *Nature (London),* 169, 495, 1952.
85. **Hunkapiller, M. W. and Hood, L. E.,** New protein sequenator with increased sensitivity, *Science,* 207, 523, 1980.
86. **Lai, C. Y.,** Regeneration of amino acids from anilinothiazolinones, *Meth. Enzymol.,* 47, 369, 1977.
87. **Lu, H. S., Yuan, P. M., Talent, J. M., and Gracy, R. W.,** A simple, rapid, manual microsequencing procedure, *Anal. Biochem.,* 110, 159, 1981.
88. See current catalogs of Pierce Chemical Company.
89. **Chang, J. Y., Braver D., and Wittman-Liebold, B.,** Microsequence analysis of peptides and proteins using DABITIC and PITC double coupling methods, *FEBS Lett.,* 93, 205, 1978.
90. **Hughes, G. J., Winterhalter, K. H., Lutz, H., and Wilson, K. J.,** Microsequence analysis III. Automatic solid-phase sequencing using DABITIC, *FEBS Lett.,* 108, 92, 1979.
91. **Silver, J.,** Microsequence analysis in automatic spinning cup sequenators, *Meth. Enzymol.,* 47, 247, 1977.
92. **Laursen, R. A.,** Solid-phase Edman degradation, an automatic peptide sequencer, *Eur. J. Biochem.,* 20, 89, 1971.

93. **Laursen, R. A.,** *Immobilized Enzymes, Antigens, Antibodies, and Peptides,* Weetall, H. H., Ed., Marcel Dekker, New York, 1975, 567.
94. **Laursen, R. A. Horn, M. J., and Bonner, A. G.,** Solid-phase edman degradation, the use of p-phenyldiisothiocyanate to attach lysine and arginine containing peptides to insoluble resins, *FEBS Lett.,* 21, 67, 1972.
95. **Horn, J. J. and Laursen, R. A.,** Solid-phase degradation attachment of carboxyl-terminal homoserine peptides to an insoluble resin, *FEBS Lett.,* 36, 285, 1973.
96. **Laursen, R. A.,** *Recent Developments in the Chemical Study of Protein Structures,* Previero, A., Pechere, J. F., and Coletti-Previero, M-A., Eds., Inserm, Paris, 1971, 11.
97. **Laursen, R. A.,** An isotope delution procedure and quantitative estimation of phenylthiohydantoins released in the Edman degradation, *Biochem. Biophys. Res. Commun.,* 37, 663, 1969.
98. **Laursen, R. A., Bonner, A. G., and Horn, M. V.,** *Instrumentation in Amino Acid Sequence Analysis,* Pesham, R. N., Ed., Academic Press, New York, 1975, 73.
99. Sequemat Solid Phase Sequencing Manual: "Sequemat Mini-15 Instruction Manual", Sequemat Inc., 1977 and "Coupling Procedures, Resin Preparations, and Miscellaneous Information on Solid Phase Sequence Analysis", Sequemat Inc., 1975.
100. **Zimmerman, C. L., Appella, C., and Pisano, J. J.,** Rapid analysis of amino acid phenylthiohydantoins by high-performance chromatography, *Anal. Biochem.,* 77, 569, 1977.
101. **Zeeuws, R. and Strosberg, A. D.,** The use of methanol in high-performance liquid chromatography of phenylthiohydantoin-amino acids, *FEBS Lett.,* 85, 68, 1975.
102. **Bhown, A. S., Mole, J. E., Weissinger, A., and Bennett, J. C.,** Methanol solvent system for rapid analysis of phenylthiohydantoin amino acids by high-pressure liquid chromatography, *J. Chromatogr.,* 148, 532, 1978.
103. **Johnson, N. D., Hunkapiller, M. W., and Hood, L. E.,** Analysis of pth-amino acids by HPLC on Dupont zorbax CN colums, *Anal. Biochem.,* 100, 335, 1979.
104. **Snapka, R. M., Sawyer, T. H., Barton, R. A., and Gracy, R. W.,** Comparison of the electrophoretic properties of triosephosphate isomerases of various tissues and species, *Comp. Biochem. Physiol.,* 49B, 733, 1974.
105. **Yuan, P. M., Dewan, R. N., Zaun, M., Thompson, R. E., and Gracy, R. W.,** Isolation and characterization of triosephosphate isomerase isozymes from human placenta, *Arch. Biochem. Biophys.,* 198, 42, 1979.
106. **Skala-Rubinson, H., Vibert, M., and Dreyfus, J. C.,** Electrophoretic modifications of three enzymes in extracts of human and bovine lens, *Clin. Chem. Acta,* 70, 385, 1976.
107. **Banner, D. W., Bloomer, A. C., Petsko, G. A., Phillips, D. C., Pogson, C. I., Wilson, I. A., Corran, P. H., Furth, A. J., Milman, J. D., Offord, R. E., Priddle, J. D., and Waley, S. G.,** Structure of chicken muscle triose phosphate isomerase determined crystallographically at 2.5 A resolution, *Nature (London),* 255, 609, 1975.

Chapter 3

ERROR MEASUREMENT METHODS IN AGING RESEARCH

Gerald P. Hirsch

TABLE OF CONTENTS

I. Detection of Altered Proteins in Aging Research 36

II. Amino Acid Misincorporation 36

III. Amino Acid Analogue Incorporation 44
- A. Heat Labile Enzymes 46
- B. Antibody-Enzyme Ratios (Cross-Reacting Materials) 47
- C. Electrophoresis and Isoelectric Focusing 48
- D. Molecular Specificity (In Vitro) 49
- E. Post-Transcriptional, Post-Translational, and Other Modifications 50
- F. Models for Error Induction 51

References 52

I. DETECTION OF ALTERED PROTEINS IN AGING RESEARCH

This chapter describes general methods used to detect altered proteins during aging and emphasizes the direct measurement of errors by amino acid misincorporation and the use of amino acid analogues for estimating fidelity. Because of the potential for mutations to be mistaken for errors in the detection of altered proteins, a general discussion of this problem precedes the discussion of the methods. This will allow the reader to consider the validity of the methods described as they relate to the error theory of aging[1,2] as opposed to the somatic mutation theory.[3]

Altered proteins are expected to arise during aging as a result of errors or mutations above some spontaneous level that occurs in young animals.[4] Mutations are distinguished from errors because they are fixed and thereby inherited from cell to cell and generation to generation. Errors, on the other hand, may appear to be inherited depending on whether the errors induced sustain a new frequency of altered proteins that is maintained or whether mechanisms exist that maintain the level of errors at some distinctive and fixed value, where the level of errors may decline after being artificially raised.[5]

The potential for effects on and interactions with one another of errors and mutations is illustrated in Figure 1. Cells which regularly divide should be more sensitive to the induction of mutations as a result of increased errors because the number of bases copied is much higher than the number replaced by repair processes in nondividing cells. Table 1 lists some of the RNAs and proteins that might contribute to errors, where mutations in the DNA coding for these molecules are likely to produce a change in the accuracy of those components that informationally feed back into the system.

The relative contribution of these potential sources can be examined in vitro after mutagenesis. For example, tRNAs from control and mutagen-treated animals can be used as substrates for mischarging by normal tRNA ligases (attachment of valine to leucine tRNA) or mischarging can be compared between ligases from control and mutagen-treated sources. At the present time, such data are not available for mammals due, to a great extent, to the lack of a definitive model for mutation induction in these species. This and other aspects of detecting altered proteins in mammals have not been investigated — rather, evidence for an increase in altered proteins has been derived from the application of model bacterial systems to mammalian aging. In most cases, changes in altered proteins during aging have been attributed to error mechanisms alone and not to potential mutational sources.[6-8]

Base substitution mutations produce almost the same classes of altered proteins as do errors. When looking at altered proteins derived from millions of cells, mutations can be distinguished from errors by special procedures such as sequencing highly purified proteins and comparing misincorporation for substitutions that can occur by base substitution mutations to those that can occur only as a result of errors.[4] If it can be shown that the level of errors, as low as 10 per million per amino acid position for each noncoded amino acid, is much more than the level of mutations, estimated to be 10^{-8}/cell division,[9] then indirect methods for detecting altered proteins would apply to the error theory of aging primarily and have little relevance to the somatic mutation theory.

II. AMINO ACID MISINCORPORATION

For many years, the focus of research on detecting altered proteins has been to test the error theory of aging. The error theory was directed primarily at the mechanisms by which preexisting errors in the components of the protein synthetic apparatus cause an increase in the level of errors in subsequent products. The original theory[1] predicted

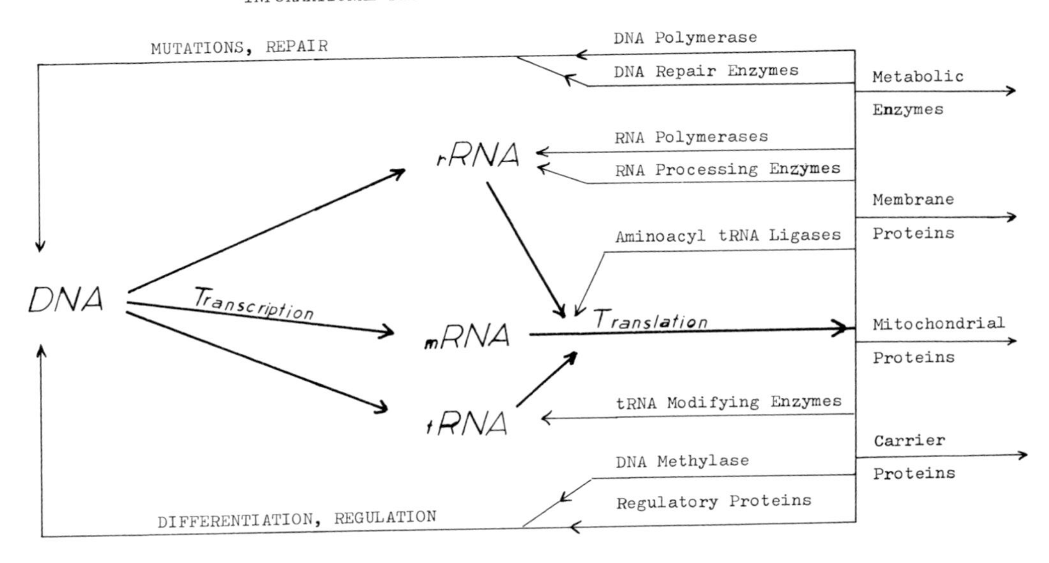

FIGURE 1. Major pathways of information flow in dividing and nondividing cells are shown by arrows. Only some pathways lead back to affect earlier steps in the general flow of information from the DNA through the messenger RNA to proteins.

Table 1
POTENTIAL RESULTS OF MUTATIONS IN INFORMATION FIDELITY

Component	Effects
tRNA	Miscoding, mischarging
rRNA	Improper processing, lowered binding of ribosomal proteins
DNA polymerase	Mutations
RNA polymerases	Errors in messages, other RNAs
DNA repair enzymes	Faulty repair, mutations
tRNA modifying enzymes	Miscoding, mischarging
RNA processing enzymes	Improper cleavage, structural changes
Amino acyl tRNA ligases	Mischarging
Ribosomal proteins	Altered structures, mispairing
Regulatory proteins	De-differentiation, re-differentiation, derepression

that errors would increase exponentially (and not be reversible), but some experimental data suggest that this may not be the case. A direct test of this theory involves the measurement of incorporation of amino acids that are not coded for in the DNA. This requires the detection of a small amount of one amino acid that should not be present in a protein or peptide and that the protein or peptide be very highly purified. This approach was applied to peptide fragments of the purified rabbit beta chain and showed misincorporation at the level of 2×10^{-4} to 6×10^{-4} for isoleucine replacing other amino acids.[10] Isoleucine was selected because a model for the recovery of the isoleucine-containing peptide was available using human hemoglobin, which has coded isoleucine in a peptide where the rabbit has valine. This approach avoided a major problem in the measurement of errors by amino acid misincorporation, namely, that one knows that the purification procedure recovers the error-containing molecule. Often the purification steps used for the removal of contaminating proteins will separate the substituted protein from the population of normal ones. A potential complication of the method using peptide fragments is the creation of new classes of contaminating peptide fragments due to nonspecific cleavage by the enzyme used to break up the purified proteins and fragments created from the presence of normal cleavage sites at unusual locations due to the misincorporation of the sensitive amino acids at various places in the protein. These problems may account for the approximate 10-fold higher value obtained in the rabbit peptide with isoleucine than the value measured as described later for isoleucine in the intact rabbit beta chain.

Contaminating fragments are not a problem when unique proteins are used for quantitating misincorporation, which lack genetic coding for one of the 20 naturally occurring amino acids. In these cases, the presence of noncoded amino acid anywhere in the protein represents misincorporation. The total amount of the noncoded amino acid divided by the total amount of all amino acids gives the average frequency of misincorporation.[11]

The measurement of errors by misincorporation requires that the protein be highly purified. The progress of purification can be followed in these proteins that lack an amino acid by labeling the protein being synthesized with the missing amino acid. Using a second labeled amino acid that is coded for, in an appropriate amount, allows for a direct comparison of the level of errors during purification steps. This general approach is illustrated in Figures 2 to 5 where marmoset hemoglobin was labeled in reticulocytes in vitro with ^{14}C-leucine (coded) and ^{3}H-isoleucine (missing).[12] After the reticulocytes were washed and the stroma removed by centrifugation, the lysate was chromatographed on Sephadex® G-200. The elution profile shows that the hemoglobin elutes between fractions 28 and 40 where the leucine label is detected. In this case, hemoglobin is the only major protein being manufactured, and it constitutes more than 90% of the total. The tritium label showing the location of isoleucine elutes throughout the profile with some indication that more of the low molecular weight protein fraction corresponds to the hemoglobin. The hemoglobin from the molecular sieve column was then chromatographed in the carbonmonoxy form on carboxymethyl cellulose. At this stage, about half of the total counts of isoleucine elute before the main fraction of hemoglobin, and there is a good correspondence between the leucine label and the isoleucine label. From the values preceding the hemoglobin peak and those that follow, contaminating proteins constitute about 1/3 of the total isoleucine label at this stage. In the following purification step (Figure 6), the hemoglobin was converted to the methemoglobin form and after concentration by pressure dialysis was rechromatographed on the same resin with the same gradient. At this step, most of the isoleucine chromatographs with the hemoglobin, except that some label occurs just prior to the first peak which consists of nonconverted carbonmonoxy or metcyan hemoglobin, and which eluted at the same pH as in the first column of carboxymethyl cellulose. In the last step, the concentrated methemoglobin was converted to metcyan hemoglobin and chromatographed a third time in the same system. In the metcyan form, the hemoglobin elutes at the same pH as it did in the first run, but now there is no indication of contaminating proteins. There is a slight shift in the peak of the isoleucine-containing hemoglobin molecules as compared to the total population. This could result from a subpopulation consisting of isoleucine substitution for charged amino acids when one chain complexed in four has a single charge difference. But there is at this stage no indication of a significant population of contaminating proteins.

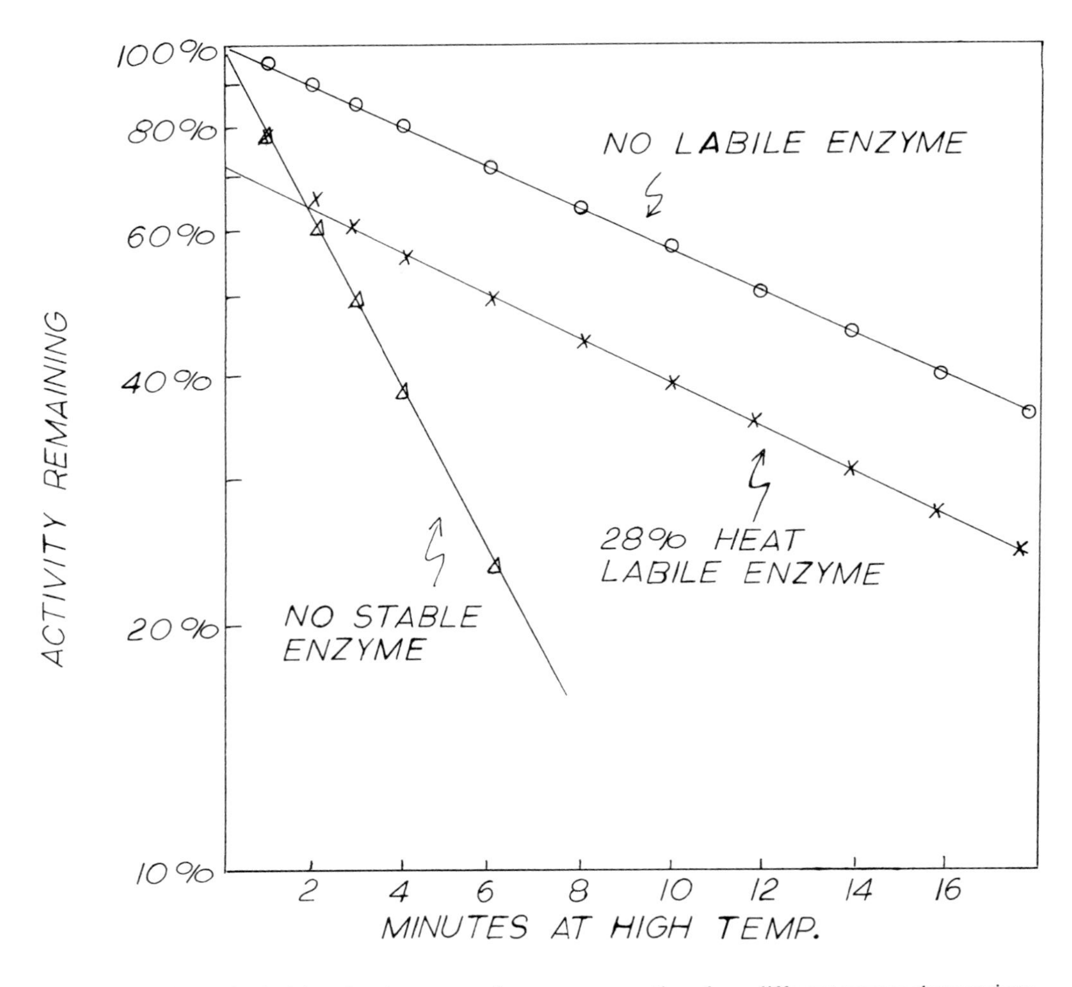

FIGURE 2. Hypothetical data for three cases of enzyme preparations from different sources show various contents of heat sensitive components. Open circles show assay points representative of most enzyme preparations. The extrapolation of data points passes through the origin with no indication of an extrasensitive subfraction. Data points labeled with "X" extrapolate to a zero-time value less than 100%. The difference represents a heat sensitive subfraction. Where the level of error becomes very high all enzymes become more heat sensitive as shown by open triangles.

A serious problem arises when using radioactive precursors to estimate errors in proteins lacking a particular amino acid. This involves the determination of specific activity of the missing amino acid. One cannot rely on the specific activity of that amino acid among other proteins because this value depends on the average turnover rate of proteins. This problem can be solved by using the specific activity ratio of the "coded" and "missing" amino acids in the protein used for error measurement and another protein that has the same amino acids both coded. Here the specific activity ratio of the amino acid coded for in both proteins is used to adjust the specific activity of the missing amino acid in the protein used to measure errors. The validity of this approach was tested using mouse hemoglobin and proteins that eluted in the high molecular weight fraction of a molecular sieve chromatography. In the mouse, both isoleucine and phenylalanine are coded for in hemoglobin and, of course, the nonhemoglobin protein fraction.[12] The specific activity ratio of these two amino acids was the same in the hemoglobin as in the other protein fraction (while the absolute values were not the same).

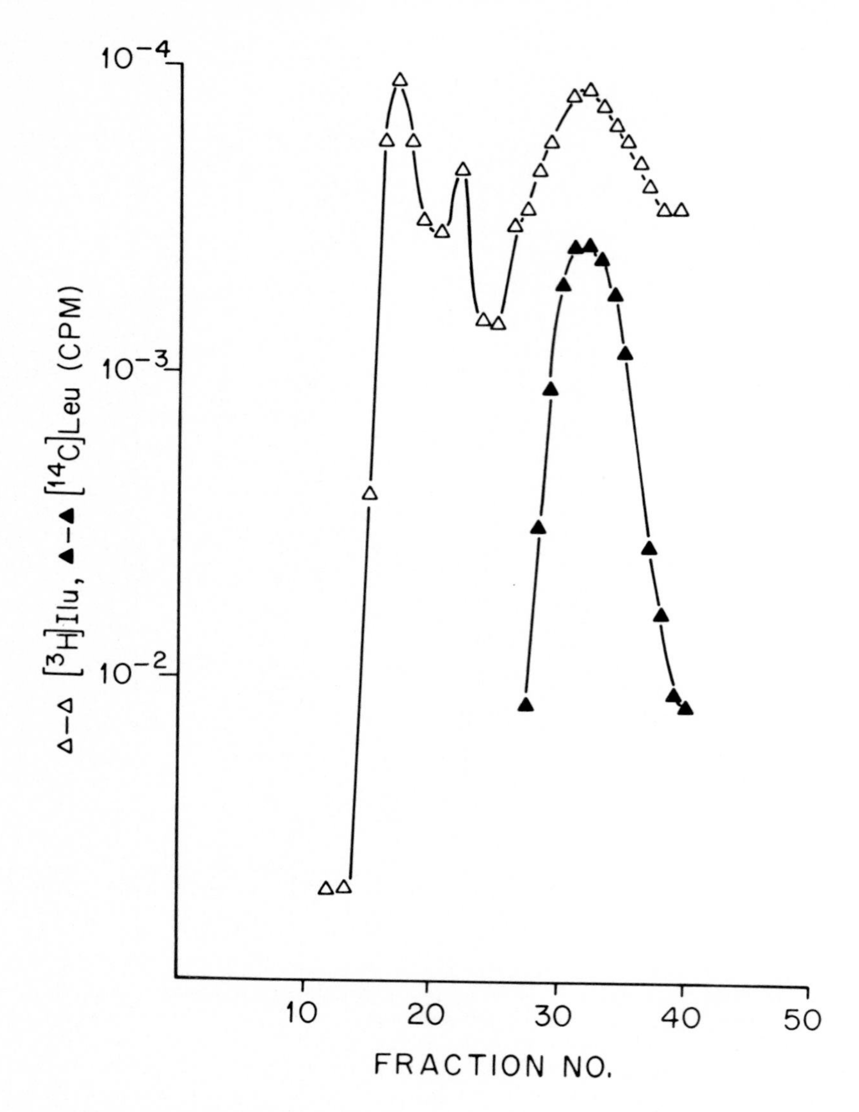

FIGURE 3. Chromatographic profile of marmoset red cell proteins, primarily hemoglobin. Mildly anemic marmoset periferal blood cells were labeled in vitro with tritiated isoleucine and ^{14}C-leucine. White cells were separated by low speed centrifugation and the red cells washed 3 times with saline. After lysis and removal of stroma the lysate was treated with carbon monoxide and chromatographed on Sephadex® G-200, as shown. The primary hemoglobin product containing leucine (closed triangles) elutes in fractions 26-38, while the majority of proteins containing isoleucine (open triangles), absent from marmoset hemoglobin, elutes in other regions where proteins of different size occur.

When the error rate is low, as it is in hemoglobins from several species, error rate measurements can be confounded by the purity of the labeled compound used to measure errors and the biological conversion of the compound. If the labeled preparation contains a normal amino acid in amounts greater than the error rate, most of the label present in the highly purified product will be the natural amino acid contaminant rather than the missing amino acid. If the rate of conversion of the amino acid probe is higher than misincorporation, then most of the label will be indicative of the converted amino acid. These potential sources can be evaluated by recovering the radio-

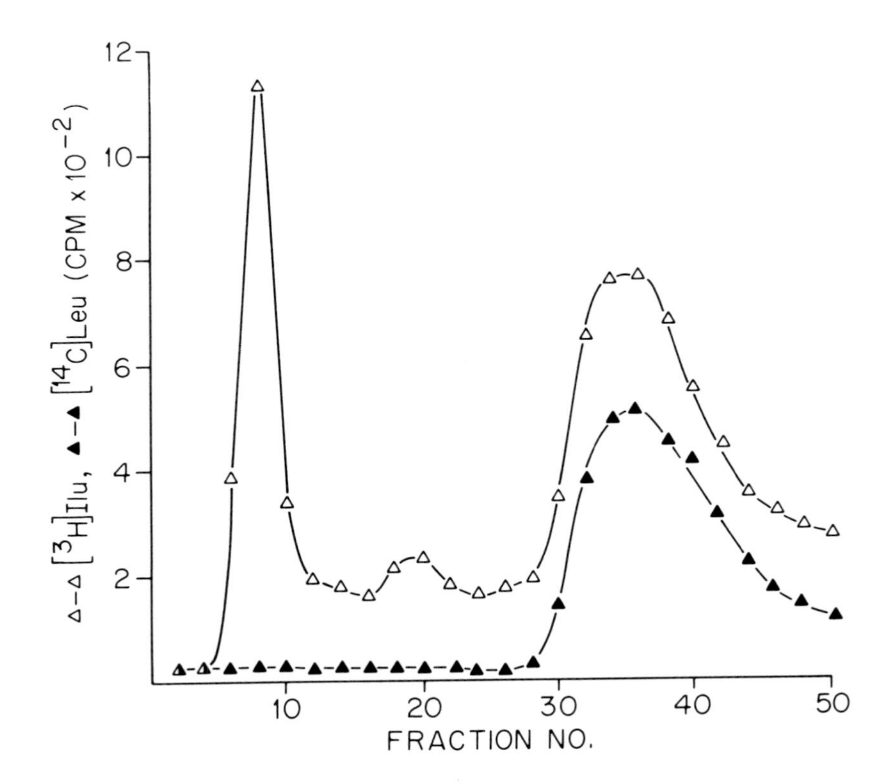

FIGURE 4. The hemoglobin fraction in Figure 3, was chromatographed on carboxymethyl cellulose using a pH gradient generated with phosphate buffers. The ^{14}C-labeled carbonmonoxy hemoglobin elutes between fractions 30 and 45 (closed triangles). Isoleucine containing (open triangles) nonhemoglobin proteins can be seen in the void volume of the column, fractions 6-10, and throughout the elution.

activity in the highly purified protein after hydrolysis and showing that it is mostly in the missing amino acid. If this is not the case, then such a recovery step will be necessary for every assay, and the profile of label of the missing amino acid in the purification schemes cannot be used to verify the purity of the product. In the case of hemoglobin in the rabbit, recovery of isoleucine was made using high pressure liquid chromatography after sequencing highly purified beta chains.[11] A significant portion of the radioactivity was recovered as isoleucine, but some label was converted to glutamic acid. This probably took place by breakdown of the isoleucine to acetate and then to the synthesis of glutamic acid.

The fidelity of hemoglobin synthesis has been examined using isoleucine misincorporation among several species.[12] Table 2 shows the average amount of misincorporation as measured by labeling of hemoglobin in reticulocytes or the quantitation of errors by amino acid analysis. The level of misincorporation is about $30/10^6$/position for most species. If a majority of error-containing molecules were degraded, the level of errors should be higher using isotopic incorporation because there has been little time for degradation of the error-containing molecules compared to time for degradation of error containing hemoglobin in mature red cells is used for analysis.

The use of methionine for the measurement of fidelity presents two additional problems that do not occur with other amino acids. The most general is use of methionine as the initiation code word in proteins. Thus, in a technical sense all proteins contain coded methionine, but in most cases the initiating methionine is removed soon after

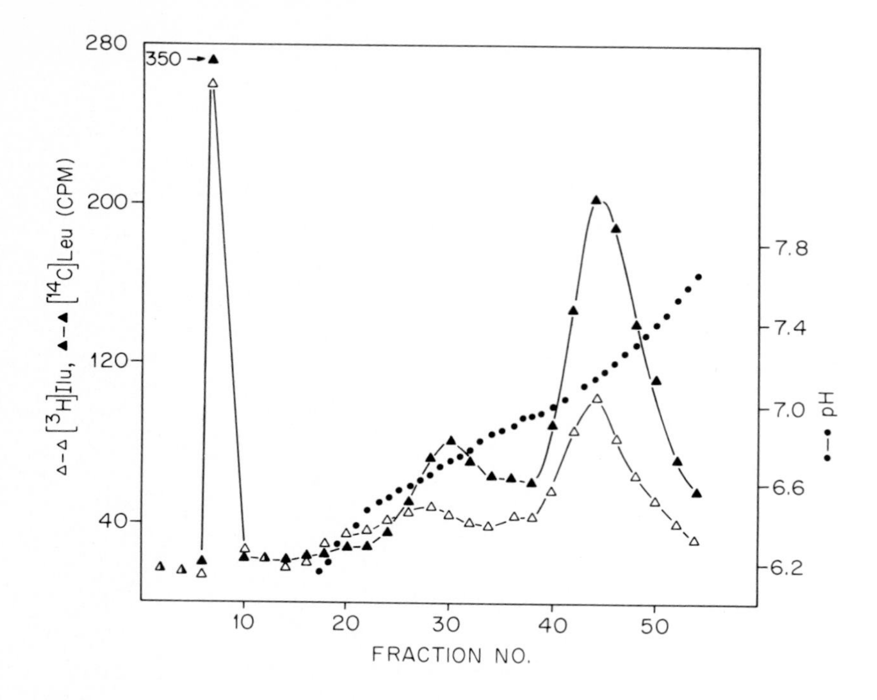

FIGURE 5. The hemoglobin protein fraction from Figure 4 was concentrated by pressure dialysis, and acidified to pH 6.0, then converted to methemoglobin with potassium ferricyanide. The methemoglobin was chromatographed under the same conditions as in Figure 4. The methemoglobin elutes at higher pH (fractions 40-50) and some unconverted carbonmonoxy hemoglobin can be seen at fraction 30. Nonhemoglobin proteins are seen at the void volume and between fractions 18 and 38. About half of the isoleucine label (open triangles) co-elutes with the leucine label hemoglobin (closed triangles).

synthesis. But, if this removal is not complete, and the residual amount of methionine at the amino terminus is greater than the amount present at other positions due to errors, then error changes during aging cannot be ascertained. The other problem results from using methionine with isotopic label in the methyl group. Because of methyl transfer reactions, the label can be moved to many other cellular components.[13]

Despite these problems, methione has been used successfully to measure errors in mouse tissue histone 1. Using sulfur-35 labeled methionine, Medvedev and Medvedva[14] measured the fidelity of histone 1, separated from histone H1° from the spleen as 10 per million. In this case, the reference proteins for specific activity were H2B and H3 plus H2A. It is likely that these histones have the same turnover rate among spleen cells, many of which are dividing cells. In other tissues which are not normally replicating, this relationship may not hold. In the thymus, methionine misincorporation was measured as five methionine errors per million per position. Further studies are needed to support the observed tissue differences in the accuracy of protein synthesis, especially because protein synthesis accuracy is thought to be an intrinsic property of the species. An alternative explanation could be invoked based on the very low level of degradation of proteins in rapidly dividing cells where error containing proteins would accumulate in dividing cells but otherwise be rapidly degraded in resting cells.[15]

The exact frequency of errors is not easily determined from proteins that require a large number of purification steps. As the purification proceeds, some of the altered molecules are discarded with the contaminants. But for comparative purposes, tightly controlled methods allow aging, tissue, and species comparisons. Some aspects of the loss of error-containing molecules can be seen in the last stages of purification or when

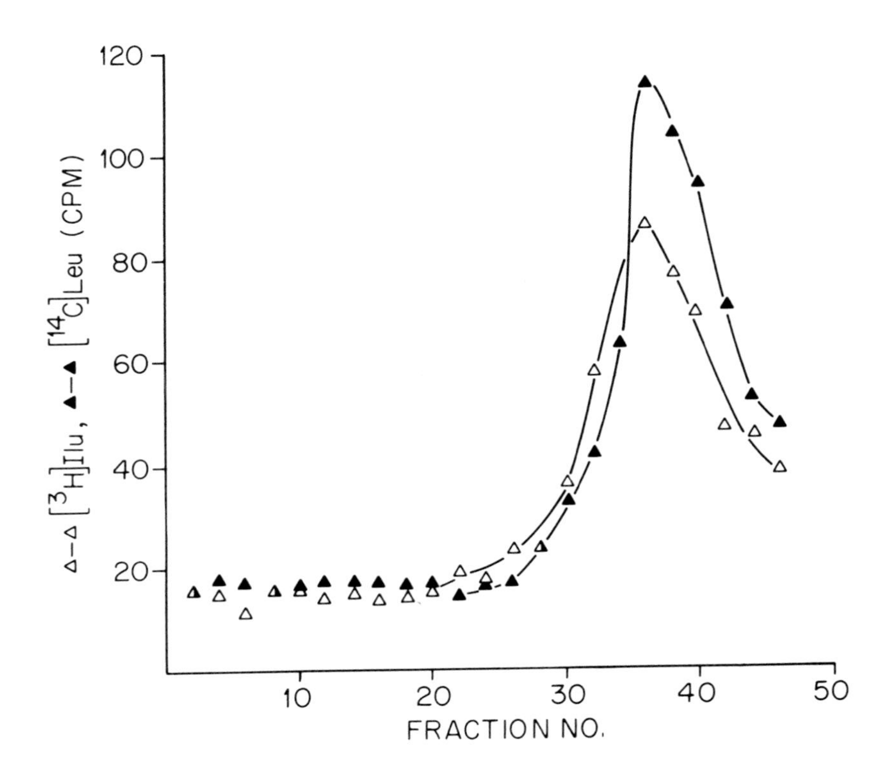

FIGURE 6. The hemoglobin fractions from Figure 5 were concentrated by pressure dialysis after conversion to metcyan hemoglobin and then acidified. As illustrated the elution profile of a third chromatography under the same conditions shows a good correspondence between "coded" leucine (closed triangles) and "noncoded" isoleucine (open triangles).

additional purification steps are used or previously used ones are repeated. The major criterion for adequate purification is that the error rate does not change when an additional step is performed, which is based on a property of the protein which was not utilized earlier (multiple molecular sieve chromatography steps will not improve the separation of a protein contaminant having the same molecular weight). Thus, the specific activity of the misincorporated amino acid should be constant. It should be kept in mind, however, that some high resolution systems that are based on the net charge of the protein will resolve proteins that differ by one charged group. If highly purified proteins are subjected to such a procedure, a subclass of the error-containing population will be separated, and it will appear that the preparation was not sufficiently pure at the previous step. Depending on the amino acid being used for the measurement of errors, a particular subclass of charge change substitutions will be included in the total group of altered proteins. The substitution frequency of isoleucine for lysine at position seven in the rabbit chain indicates that some single charge changed hemoglobin chains are recovered in this particular purification scheme.[4]

Cysteine misincorporation has been measured in *E. coli* flagellin at 6 molecules of cysteine per 10,000 molecules of flagellin.[16] Using streptomycin to increase errors, the substitution of cysteine was shown to be specific for arginine when compared to isoleucine, where arginine starvation in the presence of streptomycin caused 2.5-fold increase in misincorporation, while isoleucine starvation did change the level from that of streptomycin alone. From this and other evidence, cysteine misincorporation is thought to

Table 2
PROTEIN SYNTHESIS ACCURACY IN HEMOGLOBIN CHAINS AMONG SPECIES WITH DIFFERENT LONGEVITIES

Animal	Longevity (years)[a]	Isoleucine substitution ($\times 10^5$)	Method of analysis
Rabbit beta	13	2.4	Average of residues 1-8 from sequencing (Ref. 11)
Marmoset globin	20	40, 14, 33	Isoleucine incorporation in reticulocytes specific activity from nonhemoglobin ratio
Marmoset alpha	20	15	Amino acid analysis (donor older than 10 years)
Pig alpha	27	2.8, 3.7	Isoleucine incorporation in peripheral reticulocytes[b]
Sheep globin	20	5	Isoleucine incorporation in reticulocytes No recovery of Ilu-PTH
Cow globin	30	2.2, 3.0, 2.7	Amino acid analysis[b]
Man (N = 12) globin 20 to 60 years old	90	3.2 ± 1.5	Amino acid analysis (Ref. 24)

[a] Reference 12.
[b] Mecyan and Sephadex® G-100 chromatography steps omitted.

result only from substitution for arginine, so that the effective error rate is about 1/10,000 instead of a 20-fold lower value that would be calculated if the errors were distributed uniformly throughout the flagellin molecule.

Cysteine misincorporation in a protein of the rat seminal vesicle was found to average about 1/100,000.[17] This estimate assumes that the cysteine misincorporation is distributed throughout the protein. If in mammals the fidelity is greatly dependent on the mRNA-tRNA mispairing, as it seems to be in bacteria, this estimate of fidelity would be too low and not representative of the system as a whole. As shown for isoleucine errors by sequencing, it appears that fidelity is more uniform for at least one amino acid than for cysteine. Other sources of altered protein that may appear as misincorporation are discussed below.

III. AMINO ACID ANALOGUE INCORPORATION

The replacement of the genetically coded amino acids by other natural amino acids as errors can result from tRNA-mRNA mispairing, or mischarging of the tRNA by amino acid tRNA ligases. The incorporation of amino acid analogues, whether chemically manufactured analogous structures or amino acids that exist naturally in cells, such as homocysteine, can arise only by mischarging. If mischarging of tRNAs occurs more frequently than mispairing of tRNAs with the messenger, then the frequency of amino acid analogue incorporation should parallel errors that occur for natural amino acids. The sensitivity of amino acid analogue methods may depend on the absolute frequency of the analogue replacement. If the analogue is incorporated as rarely as natural amino acids replace one another, then the sensitivity should be about the same. If the analogue is incorporated much more frequently than natural amino acids, it is not possible to predict how the magnitude of change for the natural compounds would compare to that for the more frequently incorporated analogue. An important assumption in the use of analogues as probes for error accumulation is that the ligases responsible for the mistakes in attachment of the wrong amino acid themselves contain errors. This is not likely to be the case when analogues are incorporated frequently into proteins. As with radioactive natural amino acids used to measure misincorporation, the purity of radiolabeled amino acid analogues affects their utility for estimating errors.

Amino acid analogues have been used as a probe for estimating errors in only a few cases. Ethionine was used for aging comparisons in mouse liver among proteins and for the "ethylation" of ribosomal RNAs.[18] The radioactive tag was present in the ethyl group attached to the sulfur, so that ethionine could donate the ethyl group in reactions that would normally proceed with methyl transfer reactions and result in the methylation of proteins. Ribosomal proteins in young and old mouse liver showed different amounts of errors, depending on the size of the ribosomal protein. The aging difference ranged from nonsignificant values in small proteins to a twofold difference for higher molecular weight proteins. Selective degradation of analogue-containing molecules is the most likely explanation of the varied estimates of errors. The amount of ethylation was higher in older mouse liver, suggesting that the ability of enzymes involved in methylation reactions to discriminate against a "foreign" substrate was reduced.

The ratio of incorporation of the analogue, ethionine, to its homolog, methionine, was about 1:100.[18] It was assumed that the methionyl tRNA ligase was the primary enzyme catalyzing the incorporation of ethionine. This is likely to be the case because the error rate of ethionine incorporation if adjusted for the approximate 5% methionine content of proteins is 500 per million, more than 10 times other values for normal amino acids substituting for one another. If the methionyl tRNA ligase is the primary source for the misincorporation at a 1% frequency when error-containing proteins occur, what is the likelihood that analogue incorporation is being made by the ligases that are themselves error-containing and therefore error-prone? Using the estimates of natural amino acid misincorporation for the total population of proteins that contain one amino acid substitution, 1% to 5%, the relative contributions of the error-containing and "perfect" enzymes can be evaluated. If 1% of the methionyl tRNA ligases were responsible for the total amount of ethionine incorporation, then their average discrimination between methionine and ethionine would be unity, that is, they would catalyze the incorporation of the two substrates at the same rate. If this were the case, any change in the number of error-containing ligases would be directly seen as increased analogue incorporation. If, on the other hand, error-containing ligases still discriminate against ethionine at 10% of the value of "perfect" molecules, then the overall analogue incorporation would be equally dependent on the normal and error-containing enzymes when the population of error-containing enzymes was 5% of the total. This issue of the sensitivity of analogue incorporation to the existing population of error-containing proteins can only be resolved by further experimentation. Ethionine incorporation has been used to show increased errors in very late passage fibroblasts.[19]

An amino acid analogue that is structurally dissimilar to many of the natural amino acids is more likely to reflect the overall fidelity of amino acyl tRNA ligases. Alpha amino isobutyric acid is one such analogue that has been shown to be incorporated into protein as infrequently as natural amino acids replace one another (7×10^{-5} per position).[20] This analogue was incorporated as frequently in old animals as in young in several mouse tissues examined.[21,20] In one study, the labeled analogue was recovered after incorporation to avoid any contribution of natural amino acid contaminants (alanine is present as a low level contaminant) or products of metabolic conversion.[20] Even though this analogue was incorporated infrequently, it must still be shown that the amount of analogue incorporated will increase when the fidelity of the system is altered, showing that the fidelity measurement using analogues is dependent on the number of error-containing proteins. In vivo treatment of mice chronically with 5% dimethyl sulfoxide in drinking water affected the incorporation of analogue label in the same direction as expected.[20] However, this method of error induction was not used with recovery of label as analogue after incorporation, and further characterization of both analogue incorporation and dimethyl sulfoxide error induction is needed. While the previous discussion has focused on the need to show that analogue incorporation is a measure of preexisting errors, changes in the level of analogue incorporation during aging from other mechanisms may be equally important in understanding aging.

A. Heat Labile Enzymes

Some amino acid substitutions in enzymes alter their structure and make them less stable to denaturation at elevated temperature. This feature of altered proteins has been used to detect errors by subjecting the enzyme to a temperature that causes an exponential decline in enzyme activity over a moderate time period. When assayed at normal temperatures (25 or 37°C) after denaturation, the values of enzyme activity remaining with time extrapolate to 100% of the value determined before denaturation, for enzymes from most normal sources. However, enzyme preparations from some sources show a fraction of the total activity that is more rapidly destroyed by heat than the major portion. In some cases, several assay values were obtained for the heat-labile fraction which seems to lose activity variously with no apparent single subpopulation. In many cases no assay values were obtained that show that kinetics of denaturation of the more heat labile fraction. Figure 2 shows an idealized set of analyses for preparations with increasing levels of errors.

The loss of enzyme activity is thought to result from irreversible denaturation of the enzyme structure rather than conversion to a form which has less activity, because eventually all activity is lost. Also, no evidence has been provided that errors result in a small subpopulation of enzymes that are heat stable. A genetic heat stable variant of glucose-6-phosphate dehydrogenase has been described.[22]

The apparent heat labile enzyme activity may be lost by proteolytic activity in addition to simple denaturation. When glucose-6-phosphate dehydrogenase was heated for just 60 sec at 56°, catalytic efficiency decreased with time when assayed in lysate from mouse red cells (unpublished observations). This contrasts with constant activity assayed for nonheated lysate. If incubations are conducted for a fixed period of time after the variable heat denaturation periods, the proteolysis could represent a significant contribution to the total estimated activity loss. Kahn et al.[23] found that apparent heat labile enzyme differences disappeared after partial purification for glucose-6-phosphate dehydrogenase from human liver cells in culture.

Accurate measurement of altered enzymes using heat labile enzyme assay requires that a large number of enzyme molecules be altered. Based on estimates for the misincorporation of isoleucine[24] valine and cysteine, the frequency of substitution of noncoded amino acids is estimated to be about 30 per million per position per noncoded amino acid. For 19 potential substitutions, the rate becomes 520 per million at each position. This includes all substitutions at that position. A protein of approximately 100 amino acids (10,000 daltons) would have 5.2% of the molecules with a single incorrect amino acid somewhere in the chain, and 94.9% of the molecules would be "perfect." If only 10% of the error-containing chains result in accelerated heat denaturation (or proteolytic susceptibility), then only 0.5% of the total activity would appear as heat labile. A 10-fold increase in this level would give only 5% heat labile enzyme, and to get an accurate value at this level of enzyme loss requires that enzyme measurements be accurate to 2% of their true level. For 100,000 dalton size proteins

Table 3
STUDIES OF HEAT LABILE ENZYMES

Enzyme	Tissue	Species	Thermostable ratio (young/old)	Ref.
Glutathione reductase	Lens	Human	1.2	25
Glucose-6-phosphate dehydrogenase	Brain, spleen, liver, kidney, lung	Mouse	1.1	26
Glucose-6-phosphate dehydrogenase	Lens	Human	1.1	25
Aldolase A	Muscle	Human	1.4	27
Tyrosine amino transferase	Liver	Rat	1.0	28
Creatine kinase	Muscle	Human	1.0	27
DNA polymerase	Spleen	Mouse	1.0	29
Glucose-6-phosphate dehydrogenase	Fibroblast cultures	Human (MRC5)	1.2	30
Glucose-6-phosphate dehydrogenase	Liver culture	Human	1.4	23
Glucose-6-phosphate dehydrogenase, purified	Liver culture	Human	1.0	23
Glucose-6-phosphate dehydrogenase	Fibroblast cultures	Human (WI-38)	1.2	31
N-Acetyl beta-D-glucosaminidase	Fibroblast cultures	Human (WI-38)	1.0	31
Alpha-D-glucosidase (lysomomal)	Fibroblast cultures	Human (WI-38)	1.0	31
N-Acetyl-alpha-D-galactosaminidase	Fibroblast cultures	Human (WI-38)	1.0	31
Sulfite cytochrome C reductase (mitochondrial)	Fibroblast cultures	Human (WI-38)	1.0	31

with 10% of the substitutions conferring heat sensitivity, the sensitivity would be much better. Table 3 lists a variety of studies that have used thermal denaturation to look for errors.

B. Antibody-Enzyme Ratios (Cross-Reacting Materials)

Many mutations in the DNA coding for enzymes result in these proteins being manufactured which have no activity. Yet they maintain much of the three-dimensional structure and bind to antibodies with about the same affinity as do native proteins. This feature has been used as the basis for a method to detect altered proteins that uses antibody to precipitate the enzymes from solution. If the rate of removal of enzyme is not proportional to the loss of enzyme activity, it is assumed that in the mixture there is an excess of proteins that are structurally the same as enzymatically active molecules, but which have lost their catalytic properties. If half of the enzymes have no activity and yet the total enzyme level per cell is constant, then either twice as much protein is made in a given time period or the half-life of the enzyme must be twice as long. As seen in Table 4, the ratio of antibody necessary to precipitate 50% of the enzyme for old divided by young enzyme preparations is considerable. From 20 to 70% more antibody is necessary in some systems. The excess antibody is required because some of the proteins bind but have no enzymatic function. The parallel precipitation curves suggest that there are no subpopulations with activity but differential affinity for the antibody preparations. Cross-reacting material is not seen in all en-

Table 4
STUDIES USING ANTIBODY PRECIPITATION (CROSS-REACTING MATERIAL)

Enzyme	Tissue	Species	Cross-reacting material ratio (old/young)[a]	Ref.
Aldolase B	Liver	Mouse	1.7	32
Aldolase A	Muscle	Mouse	1.5	33
Superoxide dismutase			1.4	34
Aldolase A	Muscle	Human	1.2	27
Tyrosine amino transferase	Liver	Rat	1.0	28
Creatine kinase	Muscle	Human	0.81	27
Glucose-6-phosphate dehydrogenase	Fibroblast cultures	Human	1.04	35

[a] Ratio of antibody required for precipitation of 50% of the enzyme activity.

zymes studies and no experimental comparisons have been made between this test method and others. Estimates of protein synthesis rates and protein turnover during aging are not consistent with 1/3 change in the population of nonfunctional enzymes in old animals.

C. Electrophoresis and Isoelectric Focusing

Errors in protein synthesis necessarily result in the substitution of charged amino acids for uncharged ones or ones of net opposite charge resulting in a protein that has a net change of one or two more or less than the population of "perfect" proteins. Depending on the location of the charge change within the protein and on the net number of positive and negative amino acids, a single protein with a net change difference might be separated from the bulk population of molecules. Protein subpopulations having single charge difference were seen when their number was artificially raised. When mammalian cells in culture were starved of histidine (in the presence of histidinol), replacement occurred by the amino acid coded with the same first two codons but differing in the third (asparagine).[36] Apparently, the only substitution detected under starvation conditions involves the third position of the tRNA anticodon (wobble errors). The total percentage of protein with a single charge change seen under the starvation conditions represented 2.3% in cell lines and 10% in human fibroblasts.[37] A side band of actin seen under amino acid starvation represented a single net negative charge change of 10% of the total. While these molecules are thought to be only asparagine replacements for one of the nine histidines, sensitivity of the electrophoretic method can be evaluated by assuming that the population resulted from substitutions at most positions. This 0.1 frequency for approximately 200 amino acids (42,000 dalton actin) is equivalent to 5×10^{-4} negative 1 substitutions per position. Charge substitutions represent one-third of all substitutions so the negative subpopulation should be one-sixth. Thus, total errors should be 30×10^{-4} at this level of sensitivity, or 1.5×10^{-5} per amino acid per position. This level is more than 5 times measured misincorporation for isoleucine or methionine, so that a 2% increase in this electrophoretically separated subfraction would correspond to these levels. This subfraction is just at the limit of resolution where band spreading obscures the charge change subpopulation.[38]

Two-dimensional gel electrophoresis was applied to the separation of proteins in fruit flies, where from studies with streptomycin treatment of bacteria, single changes were known to be detectable. No population of charge changed proteins were detected, and the sensitivity of the method was estimated to be 4/10,000 (4×10^{-4}), or about 4 to 5% as much protein in a side band as in the major fraction of unaltered molecules.[5]

Table 5
STUDIES USING ELECTROPHORESIS OR ELECTROFOCUSING

Component	Tissue	Species	Age difference detected	Ref.
Superoxide dismutase	Liver	Rat	No	40
Total proteins synthesized (two-dimensional gels)	All	Fruit fly	No	5
Proteins (SDS gel electrophoresis)	Brain	Mouse	No	39
Glucose-6-phosphate dehydrogenase	Leukemic cells	Human	Yes	38

Table 6
IN VITRO METHODS

Test system	Tissue	Species	Change	Ref.
Extent of charging of tRNAs by synthetases (ligases)	Mammary tumor mammary tRNA	Human	Decreased in tumor	42
Nucleotide misincorporation with synthetic template	Fibroblasts	Human	More errors at late passage	43
Amino acyl tRNA formation and chromatography	Liver, brain	Mouse	No difference	44
Synthetic template (poly U) directed amino acid misincorporation by ribosomes	Brain	Rat	More errors in young	45
Synthetic template (poly U) directed amino acid misincorporation by microsomes	Brain	Mouse	More errors in young	46
Synthetic template (poly U) director amino acid misincorporation	Fibroblasts	Human	Late passage more accurate	47

High resolution gel electrophoresis separation by molecular weight (SDS) did not reveal any aging differences.[40] This method could easily distinguish between tissue samples from various brain regions and developing brain samples. The method was not tested for sensitivity to trace proteins and charge changed subpopulations would be poorly resolved in such a system, where the proteins are separated primarily by molecular weight, independently of charge. Isoelectric focusing was used to search for a subpopulation of superoxide dismutase protein in a purified preparation from young or old rat livers, but no such subfraction was found.[41] The isofocusing method was known to separate isozymes of other proteins, but the sensitivity of the method was not checked for the case where one population was in great excess of the other which occurs with charge changes enzymes resulting from errors.

D. Molecular Specificity (In Vitro)

The error theory of aging has had its primary focus on alterations in proteins and the accuracy with which they are made. But the concept is a general one and the scope of eventual effects pervasive for cellular functions. Thus, any cellular reaction that is highly specific is expected to show a reduced fidelity of its catalytic function if the components of its manufacture are in the first place altered. As exemplified by Table 1, a variety of targets are expected to show reduced fidelity if the overall error rate is elevated.

The contribution of fidelity of ribosomes during aging has been examined in three cases: rat liver, mouse brain, and aging fibroblasts in culture (Table 6). In each case, the synthetic homopolymer, poly U, was used to stimulate the incorporation of the proper amino acid, phenylalanine, and an "incorrect" one, leucine. The ratio of synthesis of the two amino acids is an indication of errors. In all three systems the ribosomes from old animals or late passage cells showed a lower fidelity, that is, fewer errors. The overall level of misincorporation in this system was very much higher than the in vivo systems by 10 to 1000 times. The frequency of errors is no doubt higher in these systems because it was necessary to raise the magnesium concentration in order to utilize the synthetic messages as template.

Changes in the kinds of amino acyl tRNAs were expected when purified tRNA populations were acylated with specific young or old enzyme preparations and then chromatographed to separate specific subspecies of tRNAs.[44] No differences were detected when young and old aminoacyl tRNA preparations were co-chromatographed, except for one fraction that was shown to be due to a contaminant of one isotope source. The failure to detect aging differences must be considered with regard to the sensitivity of the method to detect small amounts of altered tRNAs and very low levels of acylation with the amino acid used and tRNA of other species. The level of labeling was not sufficient to detect the trace elements of incorrect amino acylation that must occur at some low frequency.

The fidelity of DNA polymerase was tested with synthetic homopolymer templates and late passage fibroblasts were found to incorporate more of the "incorrect" base than early passage populations.[43] As with cell-free systems using synthetic templates for protein synthesis, the absolute level of "mistakes" was very much higher than would be expected based on error rates measured in vivo or mutation rates examined for whole cells, with most assays. The lowest level of misincorporation was 5/100,000, which is in the range of protein synthesis errors. In this case enzymes for late passage fibroblasts were 65 times less accurate.[43] The low level of misincorporation with polymerase from early passage fibroblasts showed that the substrate used for error detection was sufficiently pure. In this particular case, the absolute amount incorporated was actually too low to accurately measure misincorporation, so the true level may be considerably lower. In vitro bacteriophage DNA replication with natural template has an error frequency of 7/100,000 to 4/1,000,000.[9] The very high level of polymerase errors in late passage fibroblasts is thought to result in a high level of mutation detected in late passage cells using altered specificity of glucose analogues.[48] However, a comparable experiment looking for mutations in viruses that infected early or late passage fibroblasts showed no difference in the number of mutant virus particles that were produced, a frequency in the range of 3 to 4 per 100,000.[49]

E. Post-Transcriptional, Post-Translational, and Other Modifications

Altered transfer RNA species found in tumor cells and modified bases seen in tumor cells might be included in the previous section. But they can also be considered to originate from what might be considered post-transcriptional or post-translational modifications (Table 7).

Many mechanisms are known that modify the structure and function of proteins and RNAs. Ribosomal RNA is methylated, and transfer RNAs are modified in a variety of ways including various additions and some molecular rearrangements. The extent of modified bases was different in rat hepatoma transfer RNAs,[51] and some species in myeloma cells were different from normal lymphocytes.[50] Detailed examination of modified bases has not been conducted for preparation from old animals and, of course, no analysis of changes in animals treated with amino acid analogues, except ethionine, where ethylations occur where methylation additions would otherwise take place. The carcinogenic effect of ethionine is thought to involve its role in the modification of RNAs or DNA through enzymatic ethylation reactions.

Table 7
SPECIAL METHODS AND OBSERVATIONS FOR ERROR DETECTION

Method/observation (principal)	Tissue	Species	Ref.
tRNA species different in myeloma cells	Myeloma	Human	50
Modified bases in tumor cell tRNAs	Hepatoma	Rat	51
Sulfhydryl redox potential of chromatin (protein properties affected by disulfide content)	Brain, liver	Rat	52
2-Mercaptoethanol reversible changes in chromatin nuclease sensitivity (increased disulfides)	Liver	Mouse	53
Virus mutations when grown in cells (mutations caused by faulty polymerases)	Fibroblasts	Human	49
Mutations in cells (mutations caused by faulty polymerases)	Fibroblasts in culture	Human	54

A general change during aging that would be manifest as post-transcriptional modifications involves the redox potential of the cell and the resultant ratio of oxidized and reduced SH groups. An increase in the amount of disulfide bonds would change the physical and structural properties of enzymes and other cellular constituents. Age changes seen in the structure of chromatin from young and old mouse livers was reversed by treatment with 2-mercaptoethanol, a well-known reducing agent.[53] However, the quantitation of SH groups by reaction with N-ethyl malemide showed no difference during aging using chromatic substrates with brain or liver in the rat.[54] Still, the rejuvenating effects of reducing agents on immune functions in vitro[55] would suggest that some change in the redox potential occurs during aging.[56] A careful study of the levels of SH groups and SS bonds during aging is important because such a change would affect many proteins. Some drugs affect the SH redox potential of treated persons, and the therapeutic mechanism is thought to be through an increase in SH groups. Thus, it is possible to modify the SH/SS ratio in vivo and study the effect of such a change on enzyme characteristics, especially in those cases where age changes may result from such a mechanism.

F. Models for Error Induction

A clear resolution of role of errors in aging and the effect that elevated errors may have on cellular function will rely on one or more good methods for inducing errors in mammals. In bacteria, it is generally accepted that streptomycin treatment is a good way to increase errors (as much as 50 times normal), and such treatment has been successfully used to show that an artificial raising of the level of errors is not sustained when the drug is removed.[36] Rather, the level of errors reverts to the rate seen before the introduction of the drug.

Amino acid analogue feeding to animals and cells in culture has been used to induce errors. Only 2 of 12 groups of mice showed significant life shortening when fed parafluorophenyl alanine from 8×10^{-5} to 2×10^{-3} *M* in drinking water for 1 month at early age (3 to 4 weeks) or at 1 year of age.[57] There was no dose response effect, and no independent determination of error rate changes was made. Treatment of fruit flies with amino acid analogues as emerging adults for 72 hr resulted in life shortening for 5-methyl tryptophane but not ethionine or parafluorophenyl alanine.[58] Labeled ethionine was shown to have a shorter half-life after incorporation than a natural amino acid, thus showing a differential effect on function. Treatment of fruit fly larvae with amino acid analogue at late instar stages caused many of the emerging adults to die, but the survivors showed almost a normal life span. A serious problem with using fruit flies for error theory experimental procedures is the low rate of turnover of most proteins during the short life span of this species. There are only 8 periods of turnover of the average protein during the 90-day life span of these animals as compared to about 200 turnover cycles of protein synthesis components in mice.

Treatment of cell cultures with azetidine carboxylic acid, a proline analogue, for 23 hr produced 18% thermolabile glucose-6-phosphate dehydrogenase, but it also caused a four- and fivefold reduction of protein synthesis.[35] Starvation of tissue culture cells for an amino acid that produced detectable amounts of electrophoretic variants by "wobble" substitution only decreased protein synthesis to about 3% of normal.[37] There seems to be a relationship between significant error induction and protein synthesis rates. A similar correlation was seen between the rate of cell growth allowed by amino acid analogue treatment and the resulting amount of mutations detected.[59,60]

Indirect effects of the induction of errors should be anticipated. When mice were treated for 3 weeks with 5% dimethyl sulfoxide in the drinking water and then tested for an increased level of errors several confounding changes were also produced. It seemed that amino acid transport was increased, the levels of amino acids were higher and the rate of reutilization was also increased. To the extent that such changes might affect the measurements used for the quantitation of errors, incorrect conclusions might be reached. With any treatment for the induction of errors, careful attention must be paid to possible mutagenic mechanisms. Dimethyl sulfoxide might be causing mutations as well as errors, and RNA precursor base analogues might exert their effects by conversion to DNA precursors. Some amino acid analogues may affect cells by mechanisms other than changing the structure of proteins in general. As pointed out earlier, ethionine causes the ethylation of RNAs where methylation should be taking place, so the increased amount of protein synthesis seen in treated rats may not have resulted from errors directly. Other analogues such as canavanine may alter cellular function by blocking the urea cycle as opposed to a mechanism involving its incorporation into protein.

REFERENCES

1. **Orgel, L. E.**, The maintenance of the accuracy of protein synthesis and its relevance to ageing, *Proc. Natl. Acad. Sci. USA*, 49, 517, 1963.
2. **Orgel, L. E.**, The maintenance of the accuracy of protein synthesis and its relevance to ageing: a correction, *Proc. Natl. Acad. Sci. USA*, 67, 1476, 1970.
3. **Schneider, E. L., Ed.**, *The Genetics of Aging*, Plenum Press, New York, 1978.
4. **Hirsch, G. P.**, Somatic mutations and aging, in *The Genetics of Aging*, Schneider, E. L., Ed., Plenum Press, New York, 1978, 91.
5. **Parker, J., Flanagan, J., Murphy, J., and Gallant, J.**, On the accuracy of protein synthesis in *Drosophila melanogaster*, *Mech. Ageing Dev.*, 16, 127, 1981.
6. **Gershon, D.**, Current status of age altered enzymes: alternative mechanisms, *Mech. Ageing Dev.*, 9, 189, 1979.
7. **Medvedev, Z. A.**, The role of infidelity of transfer of information for the accumulation of age changes in differentiated cells, *Mech. Ageing Dev.*, 1982.
8. **Medvedev, Z. A.**, Error theories of aging, in *Alternstheorien*, Platt, D., Ed.,
9. **Hibner, U. and Alberts, B. M.**, Fidelity of DNA replication catalyzed in vitro on a natural DNA template by the T4 bacteriophage multi-enzyme complex, *Nature (London)*, 285, 300, 1980.

10. Loftfield, R. B. and Vanderjagt, D., The frequency of errors in protein biosynthesis, *Biochem. J.*, 128, 1353, 1972.
11. Popp, R. A., Hirsch, G. P., and Bradshaw, B. S., Amino acid substitution: its use in detection and analysis of genetic varients, *Genetics*, 92, s39, 1979.
12. Hirsch, G. P., Popp, R. A., Francis, M. S., Bradshaw, B. S., and Bailiff, E. B., Species comparison of protein synthesis accuracy, in *Aging, Cancer, and Cell Membranes*, Stamatoyannopoulos, G. and Neinhuis, A. W., Eds., Stratton Intercon, New York, 1980, 142.
13. Reporter, M., Methylation of basic residues in structural proteins, *Mech. Ageing Dev.*, 1, 367, 1973.
14. Medvedev, Z. A. and Medvedeva, M. N., Use of H1 histone to test the fidelity of protein biosynthesis in mouse tissues, *Biochem. Soc. Trans.*, 6, 610, 1978.
15. Shakespeare, V. and Buchanan, J. H., Increased degradation rates of protein in aging human fibroblasts and in cells treated with an amino acid analog, *Exp. Cell Res.*, 100, 1, 1976.
16. Edelmann, P. and Gallant, J., Mistranslation in *E. coli*, *Cell*, 10, 131, 1977.
17. Bradshaw, B. S., Popp, R. A., and Hirsch, G. P., Errors in protein synthesis detected by misincorporation of cysteine: effect with age and X-rays, *Fed. Proc. Fed. Am. Soc. Exp. Biol.*, 37, 1306, 1978.
18. Ogrodnik, J. P., Wulf, J. H., and Cutler, R. G., Altered protein hypothesis of mammalian ageing processes. II. Discrimination ratio of methionine versus ethionine in the synthesis of ribosomal protein and RNA of C57B1/6J mouse liver, *Exp. Gerontol.*, 10, 119, 1976.
19. Lewis, C. M. and Tarrant, G. M., Error theory and ageing in human fibroblasts, *Nature (London)*, 239, 316, 1972.
20. Hirsch, G. P., Holland, J. M., and Popp, R. A., Genetic aspects of aging, *Birth Defects Orig. Artic. Ser.*, 14, 421, 1978.
21. Hirsch, G. P., Grunder, P., and Popp, R. A., Error analysis by amino acid analog incorporation in tissues of aging mice, *Interdiscip. Top. Gerontol.*, 10, 1, 1976.
22. Kahn, A., North, M. L., Cottreau, D., Giron, G., Lang, J. M., and Oberling, F., G6PD Vientiane: a new glucose-6-phosphate dehydrogenase varient with increased stability, *Hum. Genet.*, 43, 85, 1978.
23. Kahn, A., Guillouzo, A., Cottreau, D., Marie, J., Bourel, M., Biovin, P., and Dreyfus, J.-C., Accuracy of protein synthesis and in vitro aging, *Gerontology*, 23, 174, 1977.
24. Popp, R. A., Bailiff, E. G., Hirsch, G. P., and Conrad, R. A., Errors in human hemoglobin as a function of age, *Interdiscip. Top. Gerontol.*, 9, 209, 1976.
25. Harding, J. J., Altered heat-lability of a fraction of glutathionine reductase in aging human lens, *Biochem. J.*, 134, 995, 1975.
26. Wulf, J. H. and Cutler, R. G., Altered protein hypothesis of mammalian aging processes. I. Therman stability of glucose-6-phosphate dehydrogenase in C67B1/6J mouse tissues, *Exp. Gerontol.*, 10, 101, 1975.
27. Steinhagen-Thiessen, E. and Hilz, H., The age-dependent decrease in creatine kinase and aldolase activities in human striated muscle is not caused by an accumulation of faulty proteins, *Mech. Ageing Dev.*, 5, 447, 1976.
28. Weber, A., Guguen-Guillouzo, C., Szajnert, M. F., Beck, G., and Schapira, F., Tyrosine aminotransferase in senescent rat liver, *Gerontology*, 26, 9, 1980.
29. Barton, R. W. and Yang, W-K., Low molecular weight DNA polymerase: decreased activity in spleens of old BALB/c mice, *Mech. Ageing Dev.*, 4, 123, 1975.
30. Holliday, R. and Tarrant, G. M., Altered enzymes in ageing human fibroblasts, *Nature (London)*, 238, 26, 1972.
31. Houben, A. and Remacle, J., Lysosomal and mitochondrial heat labile enzymes in ageing human fibroblasts, *Nature (London)*, 275, 59, 1978.
32. Gershon, H. and Gershon, D., Inactive enzyme molecules in aging mice: liver aldolase, *Proc. Natl. Acad. Sci. USA*, 70, 909, 1973.
33. Gershon, H. and Gershon, D., Altered enzyme molecules in senescent organisms: Mouse muscle aldolase, *Mech. Ageing Dev.*, 2, 33, 1973.
34. Reiss, U. and Gershon, D., Rat-liver superoxide dismutase, *Eur. J. Biochem.*, 63, 617, 1976.
35. Pendergrass, W. R., Martin, G. M., and Bornstein, P., Evidence contrary to the protein error hypothesis for *in vitro* senescence, *J. Cell. Physiol.*, 87, 3, 1976.
36. Parker, J., Pollard, J. W., Friensen, J. D., and Stanners, C. P., Stuttering: high-level mistranslation in animal and bacterial cells, *Proc. Natl. Acad. Sci. USA*, 73, 1091, 1978.
37. Goldstein, S., Wojtyk, R. I., Harley, C. B., Pollard, J. W., and Stannus, C. P., Fidelity of protein synthesis in aging human fibroblasts, in *Aging: Its Chemistry*, Dietz, A. A., Ed., Amer. Assoc. for Clin. Chem., Washington, D.C., 1980, 248.
38. Kahn, A., Cottreau, D., Bernard, J. F., and Boivin, P., Post-transcriptional modification of glucose-6-phosphate dehydrogenase in human leukemias, *Biomedicine*, 22, 539, 1975.

39. **Harley, C. B., Pollard, J. W., Chamberlain, J. W., Stanners, C. P., and Goldstein, S.**, Protein synthetic errors do not increase during aging of cultured human fibroblasts, *Proc. Natl. Acad. Sci. USA*, 77, 1885, 1980.
40. **Vaughn, W. J. and Calvin, M.**, Electrophoretic analysis of brain proteins from young adult and aged mice, *Gerontology*, 23, 110, 1977.
41. **Goren, P., Reznick, A. Z., Reiss, U., and Gershon, D.**, Isoelectric properties of nematode aldolase and rat liver superoxide dismutase from young and old animals, *FEBS Lett.*, 84, 83-86, 1977.
42. **Qvist, R.**, Transfer RNA and aminoacyl tRNA synthetases in hormone dependent and independent mammary tumors of GR mice. I. Comparative study of the amino acid accepting capacity of the tRNAs in the presence of the homologous and heterologous enzymes, *Can. Biochem. Biophys.*, 1, 215, 1976.
43. **Linn, S., Kairis, M., and Holliday, R.**, Decreased fidelity of DNA polymerase activity isolated from aging human fibroblasts, *Proc. Natl. Acad. Sci. USA*, 73, 2818, 1976.
44. **Frazer, J. M. and Yang, W.-K.**, Isoaccepting transfer ribonucleic acids in liver and brain of young and old $BC3F_1$ mice, *Arch. Biochem. Biophys.*, 153, 610, 1972.
45. **Ekstrom, R., Liu, D. S. H., and Richardson, A.**, Changes in brain protein synthesis during the life span of male Fisher rats, *Gerontology*, 26, 121, 1980.
46. **Kurtz, D. I.**, The effect of ageing on in vitro fidelity of translation in mouse liver, *Biochim. Biophys. Acta*, 407, 479, 1975.
47. **Wojtyk, R. I. and Goldstein, S.**, Fidelity of protein synthesis does not decline during aging of cultured human fibroblasts, *J. Cell. Physiol.*, 103, 299, 1980.
48. **Fulder, S. J. and Holliday, R.**, A rapid rise in cell varients during the senescence of populations of human fibroblasts, *Cell*, 6, 67, 1975.
49. **Holland, J. J., Kohne, D., and Doyle, M. V.**, Analysis of virus replication in ageing human fibroblast cultures, *Nature (London)*, 245, 316, 1973.
50. **Fujioka, S. and Gallo, R. C.**, Aminoacyl transfer RNA profiles in human myeloma cells, *Blood*, 38, 246, 1971.
51. **Randerath, E., Chia, L. L. S. Y., Morris, H. P., and Randerath, K.**, Transfer RNA base composition studies in Morris Hepatomas and rat liver, *Can. Res.*, 34, 643, 1974.
52. **Carter, D. B.**, No age-dependent oxidation of H3 histone, *Exp. Gerontol.*, 14, 101, 1979.
53. **Tas, S., Tam, C. F., and Walford, R. L.**, Disulfide bonds and the structure of the chromatin complex in relation to aging, *Mech. Ageing Dev.*, 12, 65, 1980.
54. **Gupta, R. S.**, Senescence of cultured human diploid fibroblasts, Are mutations responsible? *J. Cell Physiol.*, 103, 209, 1980.
55. **Kendall, P. A. and Hutchins, D.**, The effect of thiol compounds on lymphocytes stimulated in culture, *Immunology*, 35, 189, 1978.
56. **Heidrick, M. L., Albright, J. W., and Makinodan, T.**, Restoration of impaired immune functions in aging animals. IV. Action of 2-mercaptoethanol in enhancing age-related immune responsiveness, *Mech. Ageing Dev.*, 13, 367, 1980.
57. **Holliday, R. and Stevens, A.**, The effect of an amino acid analog, *p*-fluorophenylalanine on longevity of mice, *Gerontology*, 24, 417, 1978.
58. **Bozcuk, A. N.**, Testing the protein error hypothesis of ageing in Drosophila, *Exp. Gerontol.*, 11, 103, 1976.
59. **Talmud, P. J. and Lewis, D.**, The mutagenicity of amino acid analogs in Coprinus lagopus, *Genet. Res. Camb.*, 23, 47, 1974.
60. **Davies, P. J. and Perry, J. M.**, The modification of induced genetic change in yeast by an amino acid analog, *Mol. Gen. Genet.*, 162, 183, 1978.

Chapter 4

MEASUREMENT OF INTRACELLULAR PROTEIN DEGRADATION

Robert S. Bienkowski and Bruce J. Baum

TABLE OF CONTENTS

I. Introduction 56

II. Questions About Measuring Protein Degradation 56
- A. What Aspect of Protein Degradation Will Be Studied? 56
- B. What Proteins Are To Be Studied? 57
- C. What Tissue Will Be Studied? 57
- D. In Vivo or In Vitro? 57
- E. Which Isotope Will Be Used? 58

III. Measurement of Protein Degradation In Vivo 59
- A. Mathematical Preliminaries 59
- B. Example of a Labeling Experiment 63
- C. Double Label Technique 65

IV. Measurement of Protein Degradation in Cultured Cells 70
- A. Experimental Considerations 70
- B. Flux of Amino Acids Through Cultured Cells 71
- C. Calculation of Protein Degradation 72
- D. Double Label Techniques 74
- E. Applications to Aging Research 76

V. Concluding Remarks 78

Addendum 78

Acknowledgment 78

References 79

I. INTRODUCTION

The study of intracellular protein degradation has increased dramatically during the past 10 years. This review will focus on use of radioactive tracers to measure protein degradation in vivo and in vitro. It is directed at researchers who have little experience in measuring turnover, but who may consider incorporating this aspect of protein metabolism into their experiments. This is not intended to be an exhaustive review of the field; rather, our specific objectives are

1. To identify the papers which we feel are the most useful expositions of various aspects of protein turnover. Several excellent review articles are available;[1-5] in addition, clusters of papers from different laboratories contain detailed descriptions of techniques used to measure degradation in different experimental systems and are particularly valuable.
2. To pose several questions which must be considered in planning experiments to measure degradation; these questions may seem simplistic but, because they are not often asked, many studies are of marginal value, or simply wrong.
3. To discuss the mathematical aspects of turnover studies emphasizing the physical meaning of the equations.
4. To describe measurement of protein degradation both in animals and in cultured cells.

In addition to these discussions of methodologies, we have also evaluated past studies on intracellular protein degradation in aging cells. We hope that our commentary, as well as the technical descriptions provided herein, will serve to stimulate interest in examining this process in an aging context.

II. QUESTIONS ABOUT MEASURING PROTEIN DEGRADATION

In this section we discuss various practical questions about designing experiments to measure degradation. We excuse ourselves from considering the fundamental question "*What* is protein degradation?" because it has been treated in several review articles.

A. What Aspect of Protein Degradation Will Be Studied?

Cell proteins are degraded for different reasons and in various ways. The concentration of a specific protein is determined by the rates at which it is made and broken down. In the steady state, synthesis and degradation are balanced. Continual replacement of a protein may seem wasteful of energy; however, as Schimke and Doyle have observed, ability to degrade a protein rapidly correlates with ability to adjust protein levels to meet changing metabolic requirements.[1] For example, key enzymes may be regulated by varying the rates at which they are degraded. Another reason for degrading proteins is to rid the cell of functionally abnormal molecules. It is likely that the mechanism which identifies and degrades abnormal proteins differs in significant ways from the mechanism responsible for constant turnover of regulatory enzymes. Degradation occurs in different locations in the cell; there is evidence that it takes place in lysosomes as well as in cytoplasm and it is reasonable to think that the two degradation mechanisms are different. For example, lysosomal proteases are active at lower pH than cytoplasmic enzymes, and the two classes of enzymes may act on different substrates.

Clearly, one cannot conceive of protein degradation as if it were a single process and it follows that a single number will not suffice as a measure of turnover. An appropriate appreciation of events related to protein degradation requires that it be considered on several levels. The worker new to this field should approach with caution.

B. What Proteins Are To Be Studied?

The nature of the protein or group of proteins under consideration will often be an important factor in deciding how degradation will be measured. For example, is one interested in studying total protein in liver, or a specific enzyme such as catalase? Or is one interested in studying a group of liver proteins which cofractionate in some isolation procedure?

There are general rules relating physical properties to degradation (turnover) rate:[2,3] glycoproteins are broken down more rapidly than nonglycoproteins, acidic proteins more rapidly than basic proteins, and large proteins more rapidly than small proteins. These rules, together with the time frame in which an experiment is to be conducted, indicate which classes of proteins will be considered. Thus, if one is interested in studying changes in protein degradation during acute response to a drug, it would be reasonable to pay attention to proteins that are large, acidic, or glycosylated.

C. What Tissue Will Be Studied?

Characteristics of the tissue to be studied can influence how degradation will be measured. Sometimes, techniques developed for one organ cannot be used for another. All the methods we will discuss involve use of radioactive isotopes. As we will see below, (^{14}C) carbonate is a good precursor for labeling liver proteins in vivo, but is not suitable for labeling proteins in other tissues. Turnover of proteins in cultured fibroblasts can be quantitated using radioactive leucine, but phenylalanine may be a better choice when studying muscle cells.

D. In Vivo or In Vitro?

The advantages and disadvantages of using live animals or various in vitro systems such as perfused organs, organ culture or cell culture, have been discussed many times, however, it is appropriate to consider the question in the context of protein turnover experiments.

The great advantage of in vitro techniques is that the environment can be controlled precisely; these methods are very useful in studying changes in protein turnover in response to changes in nutrients and hormonal factors. It is relatively easy to approximate ideal labeling conditions in vitro (for example, constant or pulsed specific activity of the precursor can be achieved), but difficult to do this in vivo. Reutilization of radioactive precursor is probably the single most serious problem in studying turnover in vivo,[1] but this difficulty can usually be eliminated when using in vitro systems.

Organ perfusion studies have yielded valuable insights into the nature of intracellular degradation,[6-8] however, the apparatus required for these studies is expensive. Perfusion and organ culture experiments typically last a few hours and this, in fact, defines a time scale for measuring protein turnover. Only those proteins whose half-lives are comparable to the length of the experiment can be studied by these techniques; proteins with half-lives measured in days are better studied using in vivo techniques.

In using organ or cell culture, particular attention must be given to the composition of the culture medium. Incubating tissue in a balanced salt solution containing glucose and a radioactive precursor — so common in metabolic studies — is to be avoided when studying protein degradation because, under such "lean" conditions, the tissue is in a grossly catabolic state and measurements of protein turnover can be difficult to interpret.

E. Which Isotope Will Be Used?

Measurement of degradation in organ or cell culture systems is often based on release of radioactivity from tissue or cells previously incubated with a radioactive amino acid.[9,10] The validity of this technique rests on the assumption that the radioactive amino acid is not metabolized. Radioactive leucine is probably the most commonly used precursor in turnover studies. Leucine is present in significant amounts in most proteins and, because it is an essential amino acid, dilution by endogenous production is minimized. Under certain circumstances, however, leucine should not be used. For example, if the culture medium lacks certain essential nutrients, then leucine may be oxidized to CO_2.[11] In some experimental procedures, pulse labeling is achieved by incubating cells or tissue with labeled precursor for a short time and then using a chase medium containing a high level of the corresponding nonlabeled amino acid. However, when present in high concentrations leucine, and to a limited extent the other branched chain amino acids valine and isoleucine, can inhibit protein degradation in muscle; in this case, it may be preferable to use phenylalanine as a precursor.[12]

As we have mentioned, reutilization of isotope is probably the greatest technical difficulty in obtaining accurate measurements of turnover times in vivo. It has been demonstrated, for example, that as much as 50% of the free leucine pool in liver is derived from protein breakdown.[13] Amino acids released following protein degradation do not equilibrate rapidly with amino acids in the extracellular spaces, and there is evidence that the intracellular amino acid pool supplied by protein catabolism may be utilized to charge tRNA for protein synthesis.[7,14,15] Accordingly, much effort has been expended attempting to design experiments in which reutilization of the precursor is minimized. In certain cases post-translationally modified amino acids cannot be reincorporated into new protein, and their release can be used to monitor degradation. For example, production of methylated histidine (free or in small peptides) is a measure of muscle protein catabolism.[16]

Many years ago it was proposed that arginine labeled in the guanidino moiety might be a good precursor for in vivo studies of protein turnover in liver because label released upon degradative proteolysis would be cleaved by arginase present in the urea cycle (which is very active in liver).[17] Turnover times measured by following the decrease in specific activity of protein-bound arginine were found to be significantly shorter than turnover times measured using extensively reutilizable precursors such as uniformly labeled arginine or other amino acids. Nevertheless, it has been shown quite clearly that (^{14}C-*guanidino*)-arginine is reutilized because label incorporated into proteins in other tissues and released upon degradation may be taken up by liver.[18]

Perhaps the best precursor for studying protein turnover in liver is (^{14}C) carbonate. Carbonate is incorporated, via the urea cycle, into the *guanidino* group of arginine as well as the carboxyl groups of arginine and other amino acids. Turnover times of liver proteins calculated by following disappearance of (^{14}C-*guanidino*)-arginine labeled by fixation of (^{14}C) carbonate are significantly less than turnover times of proteins labeled with exogenous (^{14}C-*guanidino*)-arginine.[19] Interestingly, when carbonate labeling is employed, turnover times can be measured with equal accuracy by following the decrease in specific activity of either the (^{14}C-*guanidino*) group or the total specific activity.[19,20]

A striking example of the difference in the two labeling techniques was described by Swick and Ip in a study of liver regeneration.[19] One half of a group of rats was injected with (^{14}C-*guanidino*)-arginine and the other half was injected with (^{14}C) carbonate. One day later, half the animals in each group underwent partial hepatectomy. All animals were killed 5 days later. Compared to tissue removed at surgery, specific activity of liver proteins of hepatectomized animals labeled with (^{14}C-*guanidino*)-arginine showed no change. If the label were not reutilized, then this result would suggest

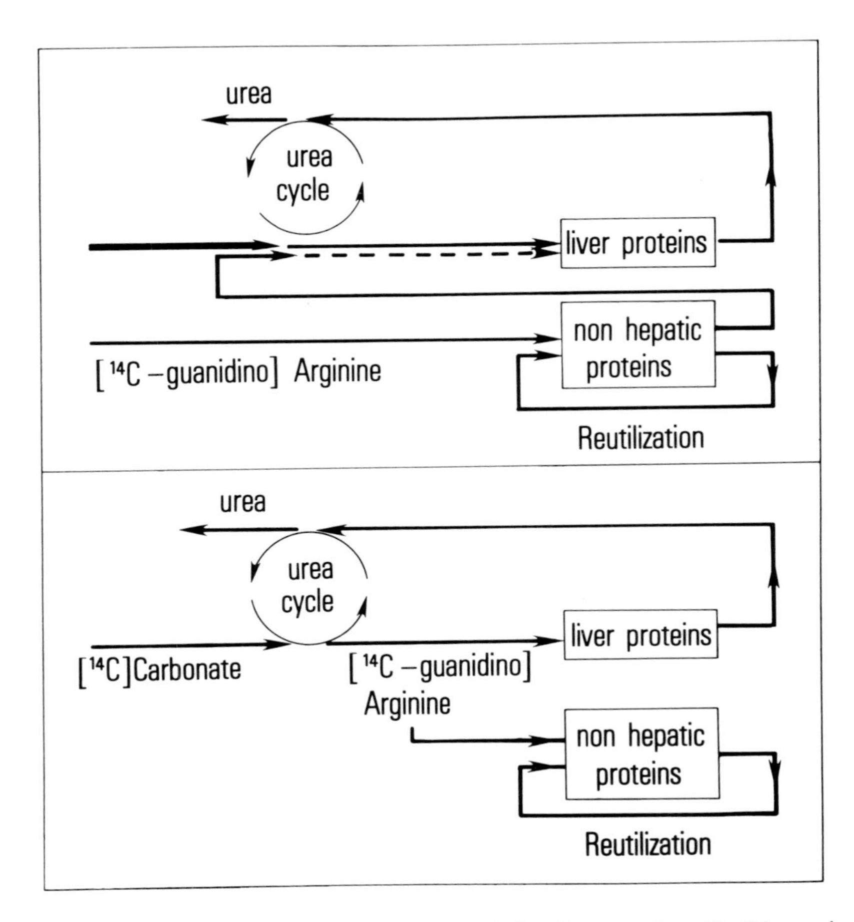

FIGURE 1. Comparison of two methods of labeling liver proteins with (^{14}C-*guanidino*)arginine. (Upper panel) The *guanidino* moiety of arginine is hydrolyzed by arginase present in liver; therefore, (^{14}C-*guanidino*)arginine labels liver proteins inefficiently. The advantage of using this tracer is that radioactive arginine released during hydrolysis of liver proteins is reutilized at only a very low level. However, breakdown of nonhepatic proteins serves as a source of free (^{14}C-*guanidino*)arginine which can be reutilized locally and also in the liver. (Lower panel) (^{14}C)Carbonate is incorporated, in the liver, into the *guanidino* moiety of arginine as well as into other amino acids. The (^{14}C) amino acids can be incorporated into proteins in the liver and other tissues. When liver proteins are degraded the released amino acids are not reutilized; upon catabolism of nonhepatic proteins free (^{14}C) amino acids can be reutilized locally, but are not incorporated efficiently into liver proteins.

that protein degradation ceased in the regenerating liver; but in hepatectomized animals labeled with (^{14}C) carbonate, the rate of protein degradation was reduced by only 40% compared to sham operated controls. However, (^{14}C) carbonate is not useful for studying turnover of nonhepatic proteins[21] because amino acids labeled in this way appear to be reutilized extensively in other tissues. The differences between the two ways of labeling hepatic protein with (^{14}C-*guanidino*)-arginine are shown schematically in Figure 1.

III. MEASUREMENT OF PROTEIN DEGRADATION IN VIVO

A. Mathematical Preliminaries

It is useful to begin with a simplified discussion of mathematical aspects of measuring protein turnover in vivo. For a more complete treatment the reader is referred to the classic works of Zilversmit.[22,23] Recently, Zak et al. have presented a very clear discussion of the subject.[24]

Let us assume we are interested in determining the half-life of a particular protein in liver. For simplicity, we consider only the case in which animals are not growing. This is a very important limitation on the range of applicability of these concepts; specifically excluded from consideration are conditions such as growth or starvation in which the metabolic state is changing with time. The protein is labeled with a radioactive amino acid delivered in an appropriate manner (e.g., single injection or continuous infusion); animals are sacrificed at various times, the protein is isolated and the specific activity (measured in units of radioactivity/mass) of the isotope is determined. In addition, the specific activity of the appropriate precursor pool (serum, aminoacyl-tRNA, free amino acid in the organ) is measured.

Consider how the specific activity of the protein can change. Since the animals are in a metabolic steady state, the sizes of the precursor and protein pools of amino acids are constant and changes in specific activity are due solely to changes in distribution of radioactivity in the pools. Let F denote specific activity of the precursor pool, and let P denote specific activity of the protein. The flow of radioactivity into protein, via synthesis, is assumed proportional to the concentration of radioactivity in the precursor pool, that is, it is described by first order kinetics:

$$(dP/dt)_{in} = K_s F \tag{1}$$

Flow of radioactivity out of protein, due to degradation, is also assumed to obey first order kinetics:

$$(dP/dt)_{out} = K_d P \tag{2}$$

As we shall see, the rate constant, K_d, in Equation 2 is related to the turnover time of the protein. Equation 2 embodies a very subtle assumption of fundamental importance for models of aging at the molecular level: given a group of *identical* protein molecules, all members of the group have equal probability of being degraded in any particular time interval. We can state this another way: molecules that have belonged to the group for a long time are not more likely to be catabolized than recently synthesized molecules.[1]

The net change in specific activity of the protein is, then:

$$\begin{aligned} dP/dt &= (dP/dt)_{in} - (dP/dt)_{out} \\ &= K_s F - K_d P \end{aligned} \tag{3}$$

In fact, the rate constants K_s and K_d are equal and Equation 3 can be written:

$$dP/dt = K(F - P) \tag{4}$$

To see why $K_d = K_s$, consider the situation in which the concentrations of radioactivity in the two compartments are the same, i.e., $F = P$. Since the protein synthesis machinery does not discriminate between labeled and nonlabeled amino acid molecules, there is no tendency, or driving force, to change the specific activity of the protein, that is $dP/dt = 0$, as predicted by Equation 4. If, however, $K_s \neq K_d$, then Equation 3 can be rewritten as:

$$dP/dt = (K_s - K_d)P \tag{5}$$

This predicts that P changes with time even though F = P which in turn, implies that the labeled amino acids are transported selectively into or out of the protein pool. This is contrary to fact and so the original assertion, namely $K_s = K_d$, is proven. Since the rate constants are independent of the specific activities, the result is valid in general even though it was established for a particular set of values.

Equation 4 is the basic equation for describing protein turnover in vivo, and it is instructive to study its solutions in some simple cases. Much of the effort that goes into measuring protein degradation is directed either to obtaining an accurate value of K for a particular protein or to determining whether different proteins are characterized by different values of K. Consider first a situation in which the protein is labeled to a specific activity P_o and then, by some artifice, the specific activity of the precursor pool is decreased instantaneously to zero. Equation 4 becomes:

$$dP/dt = -KP \qquad (6)$$

This equation describes simple exponential decay and has the solution:

$$P(t) = P_o e^{-Kt} \qquad (7)$$

In this equation t = 0 corresponds to the time at which the precursor pool is suddenly emptied of radioactivity. Note that the rate constant K has the dimension of inverse time. Indeed, one can define a *characteristic time,* 1/K, in which the specific activity of the protein declines by a factor of 1/e (= 0.368) its initial value. Radioactivity leaves the protein pool because label is released upon degradative proteolysis; therefore 1/K can also be interpreted as the time in which the number of radioactively labeled protein molecules decreases by a factor of 1/e. But, since there is nothing biologically unique about the labeled molecules, 1/K is a *turnover time* for all the protein molecules. It is often more convenient to speak in terms of the protein *half-life* which is the time in which half the number of molecules is degraded. Half-life is related to K by the equation:

$$t_{1/2} = \ln 2/K$$
$$= 0.69/K \qquad (8)$$

The half-life defines a *time scale* for an experiment. In terms of the protein of interest, a time interval, δt, is very short or very long according to whether $\delta t \ll t_{1/2}$ or $\delta t \gg t_{1/2}$. This has some practical significance, since the length of an experiment designed to measure turnover should be comparable to the half-life of the protein under study. The range of protein half-lives is enormous; values as short as 1 hr and as long as several days have been reported.[25]

Let us now consider a more realistic case in which the specific activity of the precursor pool rises very quickly (i.e., in a time $\ll$ 1/K) to a constant value F_o. Such a situation can be realized experimentally by infusing isotope into an animal at a constant rate. Equation 4 can now be rewritten:

$$dP/dt = K(F_o - P) \qquad (9)$$

This equation is often written in the form:

$$dP/dt = C - KP \qquad (9a)$$

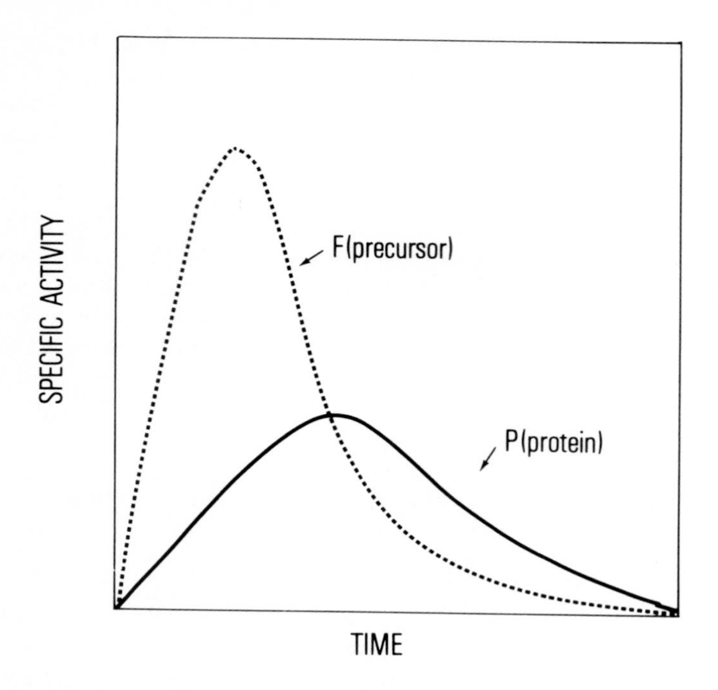

FIGURE 2. Sketch of specific activities of precursor (F) and protein (P) vs. time in an animal injected with a radioactive amino acid. Note that P always lags behind F; and that P attains maximum value (dP/dt = 0) when P = F.

in which $C = (KF_o)$ is called the zero order rate constant of synthesis and K is the first order degradation constant. Equation 9 has the solution:

$$P(t) = F_o(1 - e^{-Kt}) \tag{10}$$

If values of P and F_o are known from experiment, then Equation 10 can be solved for K. Note that the temporal behavior of P is determined by K; thus, if a protein has a very short half-life (K is large) then the steady-state value, F_o, will be approached very quickly.

When a single dose of radioactive tracer is injected into an animal, the specific activity of the precursor pool rises rapidly to a maximum value and then decreases, but not so quickly as the initial increase. Specific activity in the protein pool usually rises much more slowly than in the precursor pool. Radioactivity released from degraded protein can often re-enter the precursor pool and be reutilized for protein synthesis. The net effect of this recycling is to lengthen the apparent protein half-life. Let us return to Equation 4:

$$dP/dt = K(F - P)$$

When P is increasing, $dP/dt > 0$ and $F > P$; when P reaches a maximum value, $dP/dt = 0$ and $F = P$; and when P is decreasing, $dP/dt < 0$ and $F < P$. The general case is shown in Figure 2. Note how the specific activity of the protein pool, P, lags behind the precursor pool, F. Thus, to the extent that decay of precursor specific activity is slowed due to reutilization, protein turnover will appear to be slowed also.

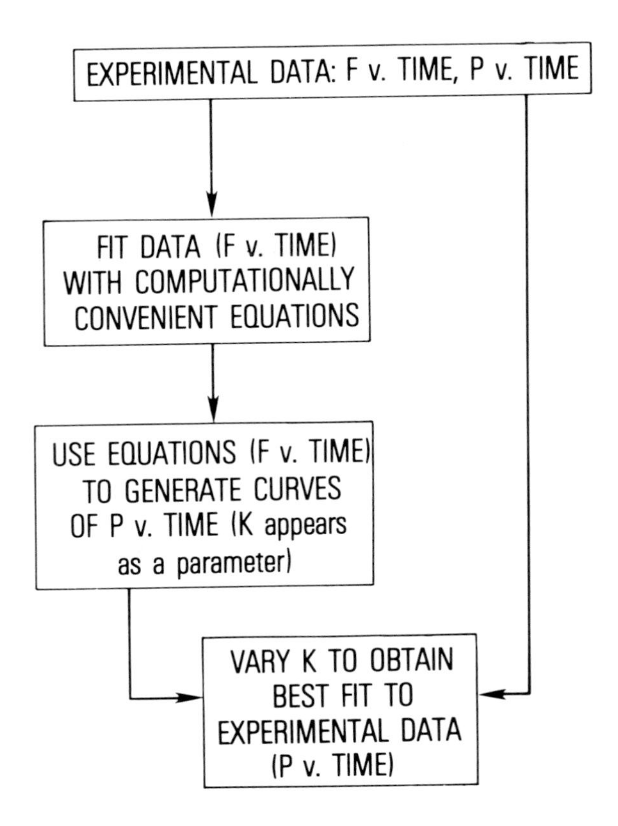

FIGURE 3. Flow chart of the operations necessary to determine protein half-life ($t_{1/2} = 0.69/K$) starting from measurements of F and P in animals injected with radioactive amino acid.

B. Example of a Labeling Experiment

To illustrate how Equation 4 can be used, and to demonstrate some of the limitations of the technique, we will consider actual data of a labeling experiment reported by Poole.[26] Before starting, it is instructive to give a brief overview of our objective. Figure 3 shows the sequence of operations which is followed in order to convert raw data into a number called the "protein half-life". Measurements of specific activity of the precursor (P) and the product (F) are made at various times after injection of isotope. The data for F v. time are fit with a curve to yield a mathematical expression for F as a function of time, F(t), Using F(t), Equation 4 is solved to generate curves of P as a function of time, P(t). In the family of curves P(t) v. time the rate constant K has the role of a parameter; its value is adjusted until the curve P(t) agrees, to within an acceptable error, with the experimental data. This particular value of K is taken as the "true" rate constant and the protein half-life is $t_{1/2} = 0.69/K$.

Poole administered single doses of (^{14}C) leucine to a group of rats and then measured specific activity of both free leucine and catalase in liver. His data show that specific activity of free leucine (assumed to be the precursor) declined rapidly during the first few hours following injection but was still measurable after 10 days. Poole fit his data with a complicated inverse function of time:

$$F(t) = \frac{16}{0.25 + t} + \frac{1.02}{(0.055 + t)^2} \qquad (11)$$

and solved Equation 4 numerically using a digital computer.

Zak et al.[24] have argued that in fitting an analytic curve to the experimental data the principal concerns should be accuracy of the fit and ease of computation; no physical meaning should necessarily be ascribed to the parameters appearing in the expression for F. Although this point of view is certainly valid, access to digital computers, while fairly easy, is not universal. Furthermore, certain expressions for F may permit analytic solutions for P which can be solved using a small calculator with a reasonable set of functions. In particular, if F can be expressed as the sum of exponential functions,[27]

$$F(t) = \sum_i f_i e^{-k_i t} \tag{12}$$

then Equation 4 can be solved for P:

$$P(t) = K \sum_i \frac{f_i}{K - k_i} (e^{-k_i t} - e^{-Kt}) \tag{13}$$

Equation 12, with 3 exponential terms, was fit to Poole's data;[26] the following expression agrees closely with his function, Equation 11.

$$F(t) = 163e^{-10t} + 26.9e^{-t} + 4.24e^{-0.1t} \tag{14}$$

The choice of decay constants ($k_1 = 10$, $k_2 = 1$, $k_3 = 0.1$ day^{-1}) is based on the assumption that the decay in F is mediated by various processes which operate on very different time scales, viz hours (or tenths of days), days and tens of days. The coefficients f_1, f_2, and f_3 were determined in the following way. At the last time point, t = 10 days, contributions from the first two exponentials (e^{-10t} and e^{-t}) are negligible, and F is determined solely by the last term, $f_3e^{-0.1t}$. From Poole's data or Equation 11, F(10) = 1.56. Therefore, the value of f_3 can be calculated:

$$\begin{aligned} f_3 &= 1.56/e^{-0.1 \times 10} \\ &= 4.24 \end{aligned} \tag{15}$$

Similarly, at t = 1 day the contribution from the first exponential is still negligible; the experimentally determined value of F(1) is 13.7. Therefore,

$$13.7 \simeq f_2 e^{-1} + 4.24e^{-0.1} \tag{16}$$

and $f_2 = 26.9$. The value of f_1 is determined in a similar way for t = 0.1 day.

Both expressions for F(t), Equation 11 and Equation 14, indicate the maximum value occurs at t = 0. Actually, F attains its maximum value a few minutes after injection; that is, the rise time for specific activity of the precursor pool is very short compared to the decay times.

Equation 13 can be used to generate curves showing the temporal evolution of specific activity for proteins of different half-lives. Figure 4 shows a family of such curves for $t_{1/2}$ = 1, 2, 4, and 8 days. Also shown is a curve for $t_{1/2}$ = 1.47 days, the half-life of catalase; the vertical bars show the ranges of values for specific activity of catalase which Poole measured.[26] Two important points can be made regarding Figure 4. If a given protein can be isolated in relatively pure form, then, using Equation 4, the half-life can be calculated from the value of K which gives the best fit to experimental data. Poole has made the important observation that experimentally derived curves of P vs. time can often be fit quite well with a single exponential function of the form $e^{-K't}$.[26]

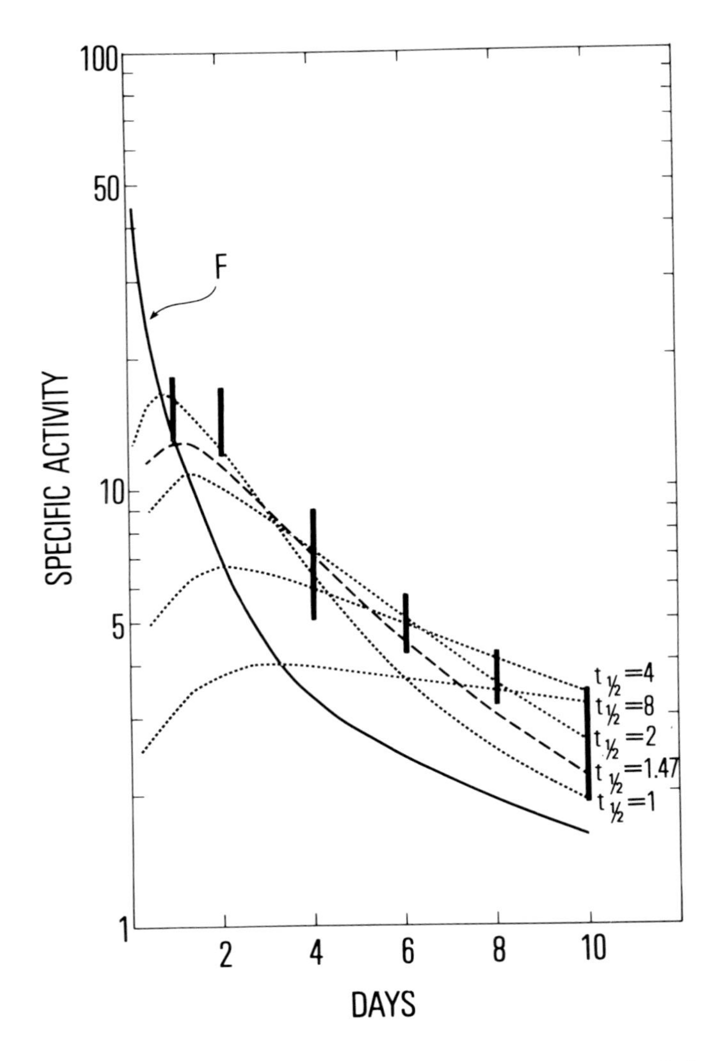

FIGURE 4. Specific activities of free (^{14}C)leucine in liver (F) and of liver proteins having various half-lives. Curve F was generated using Equation 14; protein curves were constructed using Equation 13 for different values of K. The dashed curve is for a protein with a $t_{1/2}$ = 1.47 days (half-life of catalase); the error bars are approximate representations of the range of values for catalase specific activity measured by Poole.[26]

Invariably, however, the value of K′ is less than the value of K obtained using Equation 4; that is, protein half-life is overestimated. The second important point we wish to make about Figure 4 is that for times greater than about 2 days the various curves are not very different in the sense that error bars for one curve ($t_{1/2}$ = 1.47 days) overlap several of the other curves. This means that using this technique, injection of a single pulse of radioactivity, it may be very difficult to determine whether two proteins have significantly different half-lives.

C. Double Label Technique

In 1969, Arias, Doyle, and Schimke[28] described a technique for determining whether proteins which share certain properties — for example, the proteins comprising a supramolecular structure such as the membrane of a subcellular organelle — have differ-

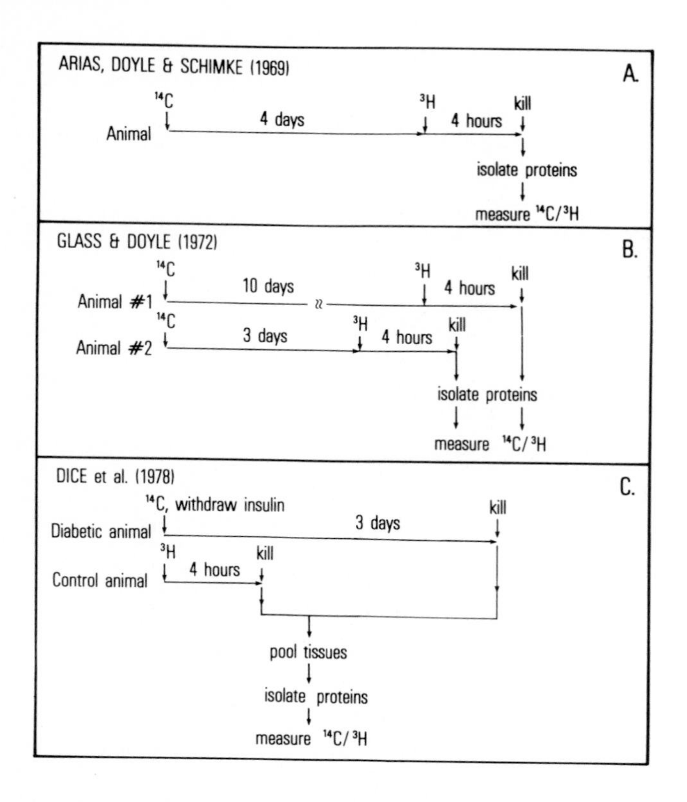

FIGURE 5. Schematic diagram showing injection schedule for various double label techniques used to study protein degradation in intact animals. (A) The original method described by Arias et al.[28] (B) A modified method used by Glass and Doyle in which degradation of short-lived proteins is studied using a 3-day interval between injections of (^{14}C) and (^{3}H) tracers, and degradation of long-lived proteins is studied using a 10-day interval.[29] (C) Method used by Dice et al. to study protein degradation when the metabolic state of the animal is altered; note that one animal is labeled with (^{14}C) and another with (^{3}H), whereas in (A) and (B) a single animal receives both isotopes.[32]

ent turnover times. In this method, an animal is injected first with a precursor such as (^{14}C) leucine; this is followed after several days with an injection of (^{3}H) leucine, and the animal is killed a few hours later. The injection and sacrifice schedule is shown diagramatically in Figure 5A. The proteins of interest are isolated and ^{3}H/^{14}C ratios determined.

At the end of the experiment the (^{3}H) proteins are still being synthesized, but (^{14}C) proteins are all in the degradation phase. The (^{14}C)-specific activity of proteins that turn over rapidly is less than the (^{14}C)-specific activity of proteins that turn over slowly. However, because proteins that are degraded rapidly are also synthesized rapidly (to maintain steady-state concentrations) the (^{3}H)-specific activity of "fast turnover" proteins is greater than the (^{3}H)-specific activity of "slow turnover" proteins. Therefore, the ^{3}H/^{14}C ratio provides a relative measure of turnover times in the sense that it is greater for "fast" proteins than for "slow" proteins. Note that because we are interested only in the ratio of the (^{3}H) and (^{14}C) radioactivities it is not necessary to determine absolute protein specific activities. An important feature of the method is that the same animal is used for both labelings. This has the effect of reducing errors due to variations between animals, however, it also constrains experimental design because the animal must be in a metabolic steady state during the course of the study. Thus, the method, as originally described, is not suitable for investigating changes in protein turnover during growth or in response to injury.

Let us consider a simple example involving two proteins having rate constants K_1 and K_2. Assume the animal can be injected with (^{14}C) leucine in such a way that the specific activity of the precursor rises instantaneously to F_C, remains constant for a few hours, t_C, and then falls to zero. Let $P_1(t_C)$ and $P_2(t_C)$ denote the specific activities of the proteins at the end of the pulse. From Equation 10 these are

$$P_1(t_C) = F_C(1 - e^{-K_1 t_C}) \tag{17a}$$

$$P_2(t_C) = F_C(1 - e^{-K_2 t_C}) \tag{17b}$$

After a time t, which is measured in days, the animal receives an injection of (^3H) leucine. Again, assume the precursor specific activity rises instantaneously to a constant value F_H and remains at this level for a few hours t_H; then the animal is killed. The specific activities of the (^3H) proteins are

$$p_1(t_H) = F_H(1 - e^{-K_1 t_H}) \tag{18a}$$

$$p_2(t_H) = F_H(1 - e^{-K_2 t_H}) \tag{18b}$$

The specific activities of the (^{14}C) proteins have been decreasing since the end of the first pulse; at the end of the experiment they are

$$P_1(t + t_H) = P_1(t_C)e^{-K_1(t+t_H)} \tag{19a}$$

$$P_2(t + t_H) = P_2(t_C)e^{-K_2(t+t_H)} \tag{19b}$$

The time dependent behaviors of the (^{14}C) and (^3H) specific activities are shown in Figure 6.

Consider, now, the ratio of (^3H) to (^{14}C) for protein 1:

$$\begin{aligned} R_1 &= (^3H/^{14}C)_{\text{protein 1}} \\ &= p_1(t_H)/P_1(t + t_H) \\ &= F_H(1 - e^{-K_1 t_H})/F_C(1 - e^{-K_1 t_C})e^{-K_1(t+t_H)} \end{aligned} \tag{20}$$

R_2 is expressed in a similar way for the second protein. In general, a pulse lasts for a very short time compared to a protein half-life. For example, in Equation 14 precursor specific activity decreases with an initial half-life of $0.69/K_1 = 0.069$ day (1.7 hr). If the protein half-lives are measured in days, then,

$$t_C, t_H \ll 1/K_1 \tag{21}$$

and one can make the following approximation:

$$e^{-K_1 t_C} \simeq 1 - K_1 t_C \tag{22}$$

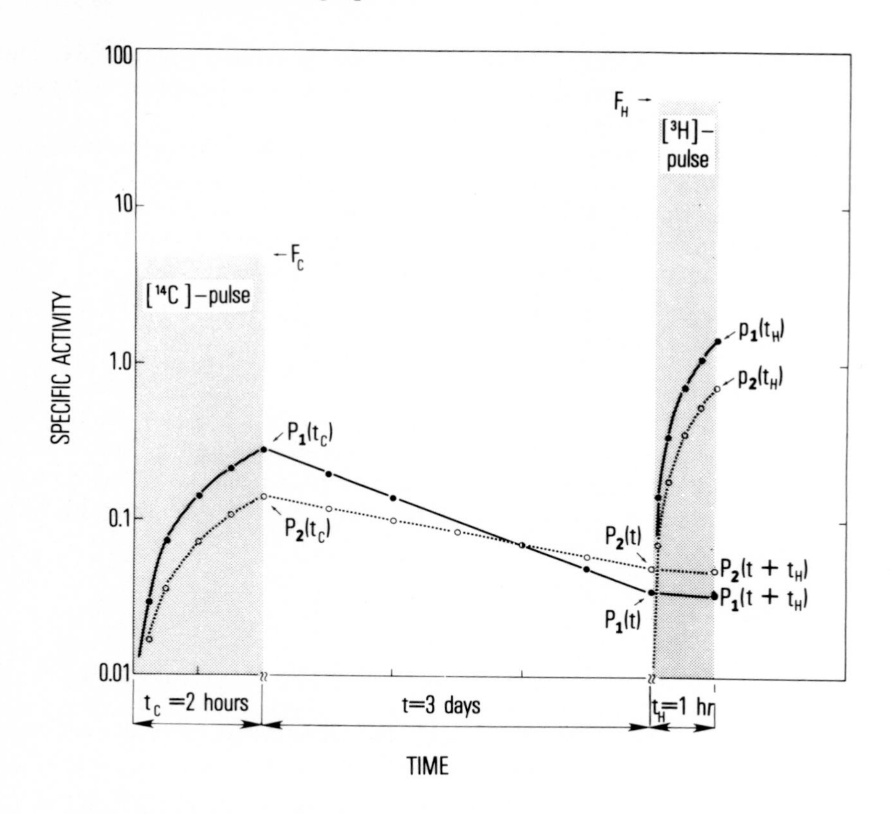

FIGURE 6. Example of a labeling experiment taking place under "ideal" conditions in which both the (^{14}C) and (^{3}H) tracers are administered as sharp pulses (shaded areas). This figure shows the changes in specific activities of two proteins with time. Protein 1 has a half-life of 1 day (K_1 = 0.69/day) and protein 2 has a half-life of 2 days (K_2 = 0.34/day). Note the changes in time scale on the horizontal axis; the scale on the vertical axis is logarithmic.

Therefore, Equation 20 becomes

$$R_1 = (F_H/F_C)\,(t_H/t_C)e^{K_1(t+t_H)} \tag{23}$$

Taking logarithms of both sides,

$$\ln R_1 = K_1(t + t_H) + \ln[(F_H/F_C)\,(t_H/t_C)] \tag{24}$$

In this idealized example, the logarithm of the isotope ratio is directly proportional to the degradation constant K_1; from Equation 8 it follows immediately that ln R is inversely proportional to protein half-life. As noted previously, the decrease in protein specific activity can often be fit quite well with a single exponential function, even when isotope reutilization is considerable. In such cases, the calculated degradation constants are less than the "true" values and one speaks of "apparent" degradation constants.

Let us now consider a case in which it is more important to know whether two proteins have different half-lives than it is to know the actual values of the degradation constants K_1 and K_2. The question arises "when is the best time to measure R_1 and R_2?" Another way of stating this is "what is the proper interval between injections of isotopes?" Clearly, this will depend on the protein half-lives. If the proteins are degraded very quickly, then there will not be much left to measure if one waits too long; however, if the half-lives are very long, then the measurements must not be made too soon because the proteins will not have degraded very much.

It is useful to consider the ratio of R_1 and R_2, Zak et al. have termed this the sensitivity index (SI):

$$SI = R_1/R_2 \qquad (25)$$

SI is a measure of the ability to distinguish between R values for different proteins. For SI = 1, $R_1 = R_2$ and the proteins cannot be resolved on the basis of their labeling kinetics. From Equation 23,

$$SI \simeq e^{K_1(t+t_H)}/e^{K_2(t+t_H)}$$

$$= e^{(K_1-K_2)(t+t_H)} \qquad (25a)$$

It can be seen that the sensitivity index increases exponentially with time. If this were the only consideration, then it would be easier to distinguish between two proteins by waiting a long time before injecting the second isotope. Recall, however, that after a very long time the proteins labeled by the first injection may not be detectable. In Equation 25a, the quantity ($K_1 — K_2$) defines a characteristic time scale; if two proteins have similar K values, i.e., if their half-lives are not very different, then one would have to wait a very long time before the value of SI was significantly greater than 1. What does it mean to say SI must be significantly greater than 1? To answer this, we must estimate the experimental error in measuring R. Arias et al. did this by administering the (^{14}C) and (^{3}H) isotopes simultaneously, sacrificing the animals after 4 hr and then determining the R values for the proteins of interest. They found that R varied by approximately 10% about the average. Thus, the experimental error in SI can be expected to be approximately 20%. Two proteins, then, have significantly different R values — and, therefore, significantly different half-lives — if:

$$SI \gtrsim 1.2 \qquad (26)$$

It is interesting to extend this discussion by determining the smallest difference in half-lives that can be detected if the two injections are administered 4 days apart. For simplicity we assume that the half-life of protein 1 is 1.5 days ($K_1 = 0.46$); then, from Equations 25a and 26:

$$e^{(0.46-K_2)x4} > 1.2$$

Taking logarithms of both sides and rearranging,

$$0.415 > K_2$$

Thus, the half-life of the second protein must be greater than 1.66 days. In principle, then, one can detect a difference in half-lives of approximately 10% (this should be regarded as the limit of sensitivity of the method). In practice, it is usually sufficient to consider proteins as having either "short" or "long" turnover times. In liver, for example, the rapidly degraded proteins are studied using a 3- or 4-day interval between injections while the slowly turning over proteins are studied using a 10-day interval. Glass and Doyle considered this problem in detail and their labeling schemes are shown in Figure 5B.[29]

Another question to consider is how long should the second labeling period last. Clearly, it should be short enough so that the labeled proteins have not begun to break down. In arriving at Equation 25a we used the approximation given in Equation 22a. Note that Equation 25a does not depend, in any significant way, on t_H. Therefore in our idealized experiment the exact length of t_H is not important. However, Martin et al.[30,31] demonstrated, in an actual labeling experiment, that the ability of this method to distinguish relatively small differences in protein half-lives depends critically on the length of the second labeling period; in studying breakdown of muscle proteins they labeled for 10 to 30 min. Nevertheless, when dealing with proteins of an organ such as liver, other factors must be considered. For example, a large percentage of protein made by liver is secreted into the blood and, it is necessary to wait sufficient time until these newly synthesized proteins are cleared. In practice the second labeling period is often 3 to 4 hr.[28,29,32]

Dice et al. modified the original double label technique in order to study degradation under nonsteady-state conditions such as starvation or drug-induced diabetes.[32] The labeling schedule is shown in Figure 5C. Consider two sets of two animals each. All animals are initially in the same steady metabolic state. At the beginning of the experiment, one animal in each set is injected with (^{14}C) leucine and one is injected with (^{3}H) leucine. After 4 hr the (^{3}H)-animals are killed and their livers are excised and frozen. One (^{14}C)-animal is treated according to the experimental protocol, while the other is maintained under control conditions. At the end of the experiment (3 to 10 days depending on the proteins of interest), the animals are killed. Each liver is combined with a liver from a (^{3}H)-animal labeled at the beginning of the experiment. The proteins to be studied are isolated and their (^{3}H/^{14}C) ratios determined for each pair of livers. While this technique allows the considerable flexibility of studying protein degradation under nonsteady-state conditions, a significant advantage of the original method is lost: because a separate animal must be used for each isotopic labeling, there is likely to be more variability in experimental results due to differences between animals.

IV. MEASUREMENT OF PROTEIN DEGRADATION IN CULTURED CELLS

In recent years, cultured cells have been used extensively to study protein degradation. In a typical experiment, cells are pulse labeled with an appropriate amino acid and release of acid soluble radioactivity during the subsequent chase period is taken as a measure of protein degradation. We will discuss this method at some length because of its potential usefulness in aging studies. Its advantages are that it is very simple to use and requires no specialized equipment. However, the simplicity of the technique belies great dangers in not designing experiments properly and misinterpreting results. In this regard, the reader is directed to the papers of Poole and Wibo,[9] Bradley et al.,[10] and Amenta et al.,[33-35] which contain very clear and detailed presentations of methodology.

A. Experimental Considerations

Choice of tracer — The main consideration in determining which amino acid to use to label protein is that it not be metabolized by the cell; that is, one wants to be sure that released radioactivity is still in the form of the original amino acid. Leucine and valine are used most often, but phenylalanine has also been employed. These are essential amino acids. Some investigators, in an effort to increase incorporation of labeled amino acid into protein, use culture medium lacking the corresponding unlabeled amino acid. Since cells cannot synthesize it themselves, they may be forced into a state of nutritional deprivation and respond to this stress by degrading relatively unimportant proteins in an effort to supply the essential amino acid necessary for synthesis of vital "housekeeping" proteins. Thus, in experiments designed to study protein degradation, the control state may, in fact, represent a state of elevated catabolic activity.

Labeling period — Cellular proteins have a wide distribution of half-lives ranging from a few minutes to several days.[1] Turnover of short-lived proteins is studied by labeling cells for a short time, 0.5 to 2 hr, and degradation of long-lived proteins is studied by labeling cells for 24 hr or more.[9,10]

Removing the labeling medium — The amount of radioactivity released during degradation is relatively small and it is necessary to have a very low background. At the end of the labeling period, unincorporated radioactivity should be removed completely. The tenacity with which extracellular amino acids associate with the cell layer is not generally appreciated. Bradley et al.[10] showed it was necessary to rinse the cell layer 7 to 8 times with a balanced salt solution before radioactivity in the wash reached an acceptably low level.

Preventing reutilization of radioactive amino acids — Reutilization of labeled amino acids which have been released from degraded protein can occur in cultured cells and lengthen the measured half-life of proteins. The easiest way to inhibit reutilization is to use a high concentration (10 × normal) of the corresponding unlabeled amino acid in the chase medium. Interestingly, reutilization appears to be a significant problem only when long chase periods are used. Bradley et al.[10] labeled fibroblasts for 40 hr with (^{3}H)-leucine and then monitored release of radioactivity into the medium for 24 hr. Between 5 and 25 hr they found a significantly greater release when the chase medium contained 2 m*M* leucine than when it contained the normal level (0.2 m*M*). The difference was very small between 0 and 5 hr.

Poole and Wibo[9] labeled rat embryo fibroblasts for 20 hr with (^{14}C) leucine and for 1 hr with (^{3}H) leucine, and then studied the effect of puromycin on release of radioactivity during a 2-hr chase. They reasoned that if reutilization were a significant problem, then cells exposed to puromycin would release more radioactivity than unexposed controls; in fact, they found no difference. It is not clear why the reutilization pathway in rat embryo fibroblasts is not as active as it is in human fetal lung fibroblasts used by Bradley et al.[10] Dell'Orco and Guthries,[36] working with human foreskin fibroblasts labeled for several days found no difference in radioactivity released after 3 hr from cells incubated with either puromycin or 2 m*M* leucine.

As mentioned earlier, high levels of leucine may have unanticipated effects on protein synthesis and it is advisable to check this in each cell line studied. A relatively simple way to do this is to measure incorporation into protein of an unrelated essential amino acid such as (^{14}C) lysine at different concentrations of unabeled leucine.

B. Flux of Amino Acids Through Cultured Cells

Before discussing how release of radioactivity is monitored and how protein degradation is measured, it is useful to consider the flux of an essential amino acid through a cell. A hypothetical scheme is shown in Figure 7. Amino acids flow between compartments or pools. The largest pool in the system is in the extracellular space which contains the culture medium. There is general agreement on the existence of a large intracellular pool which equilibrates rapidly with the extracellular pool (the rise time for saturation of this pool is approximately 7 min).[10,37] It is often assumed that this pool is a precursor of protein synthesis, however, this view has been challenged. Evidence from several laboratories indicates tRNA is charged directly from the extracellular pool.[14,15,38] (It is reasonably certain that charged tRNA is the immediate precursor of protein synthesis, and it is now relatively easy to measure specific activities of radiolabeled amino acids charged to tRNA.[39]) Recently, Wheatley and Inglis have ad-

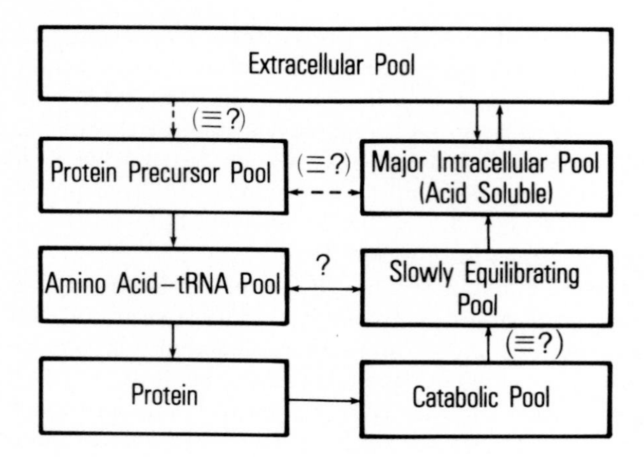

FIGURE 7. Flow of an essential amino acid through a cultured cell. A dashed arrow (- - →) means that flow between two pools is conjectured, a solid arrow (⟶) signifies that flow between two pools is probable, and the symbol (≡) means that the two pools may be identical.

vanced the hypothesis that amino acids in the large intracellular pool are not directly accessible to the protein synthesizing apparatus.[40]

Cell proteins are degraded continuously, but the fate of the released amino acids is not clear. Studies on liver — using intact animals, perfused organs, and cultured liver cells — suggest that the liberated amino acids are fed into a separate pool which exchanges very slowly with the major pool;[7,14,15,38] this second pool may also serve as a site for charging tRNA. The evidence supporting this view is that the specific activity of charged tRNA is often less than the extracellular pool but greater than the average specific activity of the intracellular pool(s). From experiments with perfused liver, Khairallah and Mortimore[7] have presented evidence that the slowly exchanging pool may be located in lysosomes.

The scheme shown in Figure 7 is based on experiments with liver. Comparable studies with cultured fibroblasts have not been carried out. However, our preliminary work with fibroblasts supports the view that at least some amino acids released from degraded protein are fed to a separate intracellular pool and that this pool exchanges slowly with the major intracellular pool and the extracellular compartment.

C. Calculation of Protein Degradation

It should be clear from the previous section that measuring release of radioactivity from cells is not necessarily the same thing as measuring release of radioactivity from degraded protein. The time course of release of radioactivity from cells to medium can be greatly influenced by kinetics of the flow of label among the various compartments.

To illustrate this, consider an experiment in which cells are labeled for 1 hr with (^{14}C) leucine, washed thoroughly and then incubated with chase medium containing 10× normal concentration of unlabeled leucine. During the chase period radioactivity in the system is redistributed among three compartments which are easily accessible for measurement. At any time, t, one can measure acid soluble radioactivity in medium, M(t); acid soluble radioactivity in cells, S(t); and acid insoluble radioactivity in cells, I(t). There is usually no significant acid insoluble radioactivity in the medium. The total radioactivity in the system remains constant during the chase; at the beginning, all the radioactivity is confined to the cell layer and $M(0) = 0$; at any other time, t:

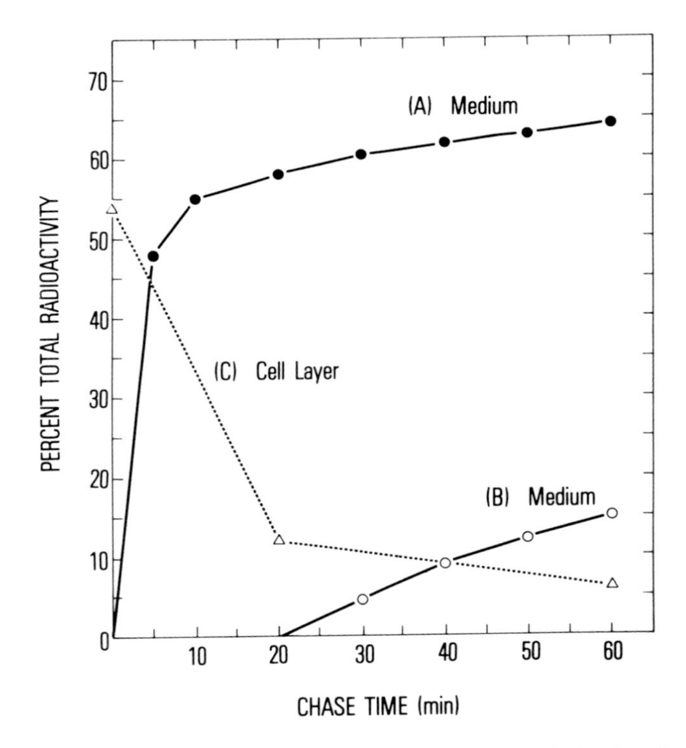

FIGURE 8. An experiment to measure protein degradation in cultured cells. Six confluent flasks of human fetal lung fibroblasts were labeled with (^{14}C) leucine for 1 hr, then the labeling medium was discarded and the cell layers were washed 8× with a balanced salt solution. Two flasks were used to determine the amounts of acid soluble and insoluble radioactivity in the cells at the beginning of the chase ($t = 0$); these quantities are designated S(0) and I(0), respectively. Two flasks were incubated with chase medium for 20 min; then the medium was discarded and S(20) and I(20) were determined (See Equation 29). The two remaining flasks were incubated for 1 hr. The chase medium was changed at $t = 5, 10, 20, 30, 40, 50$, and 60 min and acid soluble radioactivity in the medium M(t), was measured at each time; S(60) and I(60) were determined at $t = 60$ min. Curve A shows time course of acid soluble radioactivity in the medium expressed as a percentage of total radioactivity in the cell layer at the beginning of the chase [S(0) + I(0)]. Curve B shows acid soluble radioactivity in the medium expressed as a percentage of total radioactivity in cell layer after a 20 min chase [S(20) + I(20)]. Curve C shows time course of acid soluble radioactivity in the cell layer expressed as a percentage of [S(0) + I(0)].

$$S(0) + I(0) = M(t) + S(t) + I(t) \tag{27}$$

Depending on how these measurements are used, the calculated amount of degradation can vary significantly.

Figure 8 shows results of an experiment to measure protein degradation in human fetal lung fibroblasts (details of the method are given in the legend). The upper curve (A) shows release of radioactivity to the medium expressed as a percentage of total radioactivity in the system at the beginning of the chase. Note the very steep rise in the curve during the first 10 min; this represents release of radioactivity from the large intracellular pool. Sometimes, this curve is used to calculate the rate of degradation by measuring the increase in acid soluble radioactivity in the medium between times t_1 and t_2:

$$\text{Rate of degradation} = \frac{M(t_2) - M(t_1)}{S(0) + I(0)} \cdot \frac{100}{t_2 - t_1} \qquad (28)$$

Using this equation, the rate of degradation between 20 and 60 min is calculated to be 6% per hour. This is actually an underestimate because the denominator includes the acid soluble radioactivity present in the large intracellular pool at the beginning of the experiment.

In a variation of this calculation, radioactivity secreted into the medium during the first 20 min is discarded. In Figure 8, curve (B) shows acid soluble radioactivity in the medium relative to total radioactivity in the cell layer at 20 min. Instead of Equation 17 we now have, for $t > 20$ min,

$$S(20) + I(20) = M(t) + S(t) + I(t) \qquad (29)$$

Equation 28 is replaced by:

$$\text{Rate of degradation} = \frac{M(t_2) - M(t_1)}{S(20) + I(20)} \cdot \frac{100}{t_2 - t_1} \qquad (30)$$

Using this equation, the calculated rate of degradation is increased significantly to 15% per hour. However, this calculation does not take account of the possibility that appearance of radioactivity in the medium may be determined more by kinetics of intracellular pools emptying into the extracellular space than by release of radioactive amino acids from catabolized proteins. If the slowly equilibrating pool exists, then breakdown of protein with half-lives less than the turnover time of the pool would not be detected.

The most direct way to measure protein breakdown is to measure the increase in acid soluble radioactivity in the *entire system*. After allowing the major intracellular pool to empty, the medium is discarded and replaced with fresh chase medium. One set of culture flasks is incubated for time t_1 and another for time t_2. The correct equation to use for calculating degradation is

$$\text{Rate of degradation} = \frac{[M(t_2) + S(t_2)] - [M(t_1) + S(t_1)]}{[S(20) + I(20)]} \cdot \frac{100}{t_2 - t_1} \qquad (31)$$

Using this expression, the rate of degradation is 8% per hour. The similarity between numerical results obtained using Equation 28 and Equation 31 is to be regarded as fortuitous.

D. Double Label Techniques

Various methods have been designed to study protein degradation in cultured cells using double label techniques. Johnson and Kenney studied turnover of a particular enzyme, aminotransferase, in H35 cells.[41] They addressed the question whether the protein half-life changed when the enzyme was synthesized under conditions which rendered it heat labile and less active. Their strategy, shown in Figure 9A, was to label cells with (^{14}C) under control culture conditions for 2 hr and then label cells with (^{3}H) under experimental conditions for 2 hr. The enzyme synthesized during each labeling period decayed exponentially (i.e., followed first order kinetics) during the subsequent chase, and, in fact, the rate constants were the same for both cases.

Tweto and Doyle studied turnover of membrane proteins in HTC cells.[42] Their labeling schedule is shown in Figure 9B. One flask of cells was labeled for 30 min with

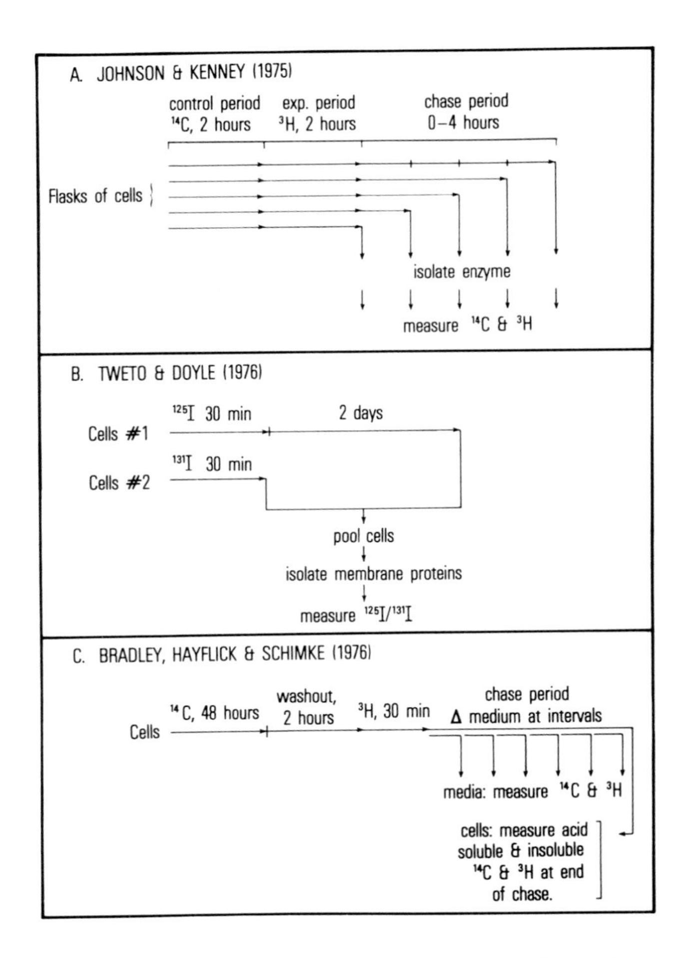

FIGURE 9. Examples of double label techniques used to study protein degradation in cultured cells. (A) Method used by Johnson and Kenney to investigate changes in protein half-life when cells are cultured under different conditions.[41] (B) Method developed by Tweto and Doyle[42] to study turnover of membrane proteins; note similarity between this method and the in vivo technique shown in Figure 5C. (C) Method of Bradley et al.[10] and Poole and Wibo;[9] note that the (^{3}H) and (^{14}C) tracers label different classes of proteins.

(^{131}I) and then frozen immediately; another flask was labeled with (^{125}I), also for 30 min, and then incubated for 2 days. At the end of the chase period, the two sets of cells were combined and membrane proteins were isolated. (^{131}I) measured relative amounts of protein present at the beginning of the experiment while (^{125}I) measured relative amounts remaining at the end. The ratio (^{125}I/^{131}I) was taken as a measure of rate of degradation relative to rate of synthesis. Note the formal similarity between this method and the method of Dice et al. for in vivo labeling shown in Figure 6C.

Bradley et al.[10] as well as Poole and Wibo,[9] have described a very different double label technique. The basis for this method is the concept that proteins synthesized at low rates are also degraded slowly, while proteins synthesized at rapid rates are also broken down quickly. Thus, one can selectively label populations of proteins having either "fast" or "slow" turnover kinetics by incubating cells for either "short" or "long" times. Operationally a short time is 0.5 to 2.0 hr, while a long time is 24 to 48 hr. Both fast and slow proteins can be studied in the same experiment; the labeling

Table 1
COMPARISON OF EXPERIMENTAL SYSTEMS USED TO STUDY PROTEIN DEGRADATION IN FIBROBLASTS IN LOW AND HIGH PASSAGE

Cell type	Culture medium	Labeling protocol	Ref.
Human fetal lung (WI-38)	Eagle's basal medium (BME)	Short time: (^{3}H)-leucine × 30 min Long time: (^{14}C)-leucine × 40 hr	10
Human fetal Lung (MRC-5)	BME	Long time: (^{3}H)-leucine × 5 days	44
Human fetal lung (MRC-5)	Eagles' minimum essential medium (MEM) and medium 199	Short time: (^{3}H)-valine × 30 min	45
Adult skin	MEM	Short time: (^{3}H)-phenylalanine × 1 hr Long time: (^{14}C)-phenylalanine × 24 hr	46

procedure is shown in Figure 9C. Note that there is a fundamental difference between this method and the methods shown in Figure 9A and B; in (A) and (B), the same proteins, or groups of proteins, are labeled with each isotope, whereas in (C) different classes of proteins are labeled.

E. Applications to Aging Research

The loss of proliferative capacity in serially passaged diploid fibroblasts has been widely used as a model of the aging process.[43] Several groups have considered the problem of how patterns of protein degradation vary during aging in vitro. While the question may seem simple, it is interesting that the results are so disparate. Part of the reason, we believe, lies in subtle differences of technique. For example, different kinds of culture medium are used. Table 1 summarizes the procedures used by investigators who have studied problems related to aging.

Some general comments should be made about certain aspects of the methods. Shakespeare and Buchanan labeled cells for 5 days in leucine-free medium.[44] As pointed out earlier, protein degradation is likely to be increased when an essential amino acid is omitted from the nutrient medium. Therefore, this report should be evaluated with particular caution. Dean and Riley used valine-free medium when labeling with (^{3}H) valine for 30 min.[45] It is not clear what effect this brief period of deprivation has on protein degradation, but it would be simple (and useful) to do an experiment to examine this question. Bradley et al.[10] used a very efficient technique, which they call ''intermittent perfusion'' to empty intracellular pools of radioactivity rapidly; Dean and Riley[45] described a washing procedure which they state was effective in depleting the major intracellular pool (however, we have been unable to duplicate this result with other fibroblast lines). Goldstein et al.[46] did not attempt to empty the large intracellular pool; instead, they tried to take account of it in their calculations of degradation. In this they were not totally successful because they include the pool when calculating total radioactivity in the system; they used Equation 28 rather than Equation 31, thereby underestimating the amount of degradation.

Because of differences in experimental design, the various reports cannot be compared directly. Here, we attempt a general comparison by posing questions (retrospectively) which can be answered by considering data presented in the original papers. Since this involves our own interpretations of the authors' results we indicate by a parenthetical reference the table or figure in the original report on which we base our conclusions.

1. Are average half-lives of short-lived proteins in early and late passage cells significantly different? Bradley et al.: yes; in senescent fibroblasts the average half-life

is 2.7 ± 0.6 hr while in early passage cells it is 5.0 ± 0.4 hr (Table 2, Ref. 10). Dean and Riley: no; in early passage the half-life is 0.88 hr while in late passage it is 1.1 hr (Table 1, Ref. 45).

Comment: There is no obvious reason why the proteins examined by Dean and Riley turn over so much faster than those studied by Bradley et al. However, the fact that Dean and Riley labeled in absence of valine may provide a possible explanation.

2. Are the half-lives of long-lived proteins in early and late passage cells significantly different? Bradley et al.: yes; the half-life in early passage cells is 58 ± 15 hr while in late passage it is 38 ± 15 (Table 1, Ref. 10). Dean and Riley: no; the average half-life for both age groups is 32 ± 3 hr (Table 2, Ref. 45).

   Comment: We are cautious in accepting the claim of Bradley et al. that the difference in half-lives is significant. Our calculation, using a two tailed t-test, yields a p value between 0.05 and 0.10. At best, this is marginally significant. The labeling period used by Dean and Riley (0.5 hr) does not label long-lived proteins efficiently. To study this class of proteins it is better to label for several hours.

3. Which cells — low or high passage — degrade more protein in a 3-hr chase period following a short labeling period? Bradley et al.: low passage (Figure 9, Ref. 10). Dean and Riley: high passage (Table 2, Ref. 4). Goldstein et al.: low passage (Table 3, Ref. 46).

   Comment: There is no obvious explanation why the result of Dean and Riley should be in disagreement with the findings of the other two groups. The major difference in technique appears to be that Dean and Riley used valine-free medium.

4. Which cells — high or low passage — degrade more of the long-lived proteins? Bradley et al.: there is no difference after a 3-hr chase (Figure 9, Ref. 10). Goldstein et al.: no difference after a 3-hr chase (Table 3, Ref. 46). Shakespeare and Buchanan: after a 2-day chase, late passage cells degrade more protein (Figure 3, Ref. 44).

   Comment: We tend to discount the report of Shakespeare and Buchanan because they cultured cells in the absence of an essential amino acid for several days. (The next 2 questions relate to use of canavanine — an analogue of arginine — in culture medium; proteins which incorporate canavanine have an abnormal conformation.)

5. When cells are labeled in presence of canavanine for a short time, how does the percent of label released during the first few hours of chase vary with passage? Bradley et al.: more is degraded in late passage (Figure 11, Ref. 10). Dean and Riley: there is little difference (Table 2, Ref. 45).

   Comment: Bradley et al. used arginine-free medium and canavanine at a concentration of 5 m*M*. Dean and Riley used complete culture medium containing a tracer amount of (^{14}C) canavanine (ca. 1 μM) and valine-free medium supplemented with a tracer amount of (^{3}H) valine (ca. 3 pM). Analogues are usually incorporated into proteins much less efficiently than the corresponding amino acids. To drive the process in favor of the analogues, one should either eliminate the amino acid from the culture medium or use a very high concentration of analogue. While Dean and Riley did show that (^{14}C) canavanine was incorporated into protein, it is not clear that the amount was sufficient to cause a significant fraction of protein molecules to have an abnormal conformation.

6. Are analogue-containing proteins degraded faster in early or late passage cells? Bradley et al.: for a given passage level, analogue-containing proteins are always degraded faster; at high passage the difference is smaller than at low passage (Table 3, Ref. 10). Dean and Riley: analogue-containing proteins are degraded faster at all passage levels; the half life does not depend on passage level (Table 1, Ref. 45).

V. CONCLUDING REMARKS

Studying normal physiological processes across the age spectrum is an especially complicated endeavor. No one metabolic event in any cell is isolated; an organism's general physiological status can be of considerable influence on cellular functions. In a young adult animal these influences may be thought of as somewhat optimal. In an aged animal this may not be the case. Such considerations are appropriate for investigations of intracellular protein degradation during aging.

Many recent studies have demonstrated that a variety of conditions and agents which can modulate this catabolic process. For example, the role of nutritional status in protein breakdown has been intensively studied in both humans and laboratory animals[16,32,47] as well as cultured cells in vitro.[9,48] A number of hormones such as glucagon,[49,50] ACTH,[51] fibroblast growth factor,[48] insulin,[48,50] and prostaglandin E_1,[52] have also been shown to affect intracellular protein degradation, and drugs as ergot alkaloids[53] and corticosteroids[48] may act as regulatory factors. Since altered organ function, systemic disease and use of therapeutic drugs (in humans) are much more common in the aged organism than in the younger counterpart, and their possible influence must be a factor in design of experimental protocols and interpretation of data.

Our purpose has been to provide the concerned investigator with the appropriate background and perspective with which to evaluate the methodologies available for meaningful study of intracellular protein degradation. Our discussion of previous reports of protein degradation during aging demonstrates that at present we are unable to make a general assessment of this process with age. Clearly more studies, in other systems, using rigidly defined conditions are necessary before this can be done. Intracellular protein degradation is an important metabolic process. Hopefully, we can soon achieve a better understanding of the status of protein degrading mechanisms in the aged organism.

ADDENDUM

We wish to direct attention to various reviews on protein degradation which have been published since the manuscript for this chapter was completed. Hershko and Ciechanover[54] summarize recent work in this field and present a lucid discussion of an energy-dependent mechanism for degrading proteins. Amenta and Brocher[55] discuss briefly, but comprehensively, mechanisms of protein turnover in cultured cells. Rannels and colleagues[56] consider problems related to use of radioactive precursors to study protein metabolism.

ACKNOWLEDGMENT

R. S. Bienkowski would like to acknowledge support from the U.S. Public Health Service (grant HL 22729) and the March of Dimes — Birth Defects Foundation.

REFERENCES

1. Schimke, R. T. and Doyle, D., Control of enzyme levels in animal tissues, *Ann. Rev. Biochem.*, 39, 930, 1970.
2. Goldberg, A. L. and Dice, J. F., Intracellular protein degradation in mammalian and bacterial cells, *Ann. Rev. Biochem.*, 43, 835, 1974.
3. Goldberg, A. L. and St. John, A. C., Intracellular protein degradation in mammalian and bacterial cells, *Ann. Rev. Biochem.*, 45, 747, 1976.
4. Doyle, D. and Tweto, J., Measurement of protein turnover in animal cells, in *Methods in Cell Biology*, Vol. X, Prescott, D. M., Ed., Academic Press, New York, 1975, 235.
5. Segal, H. L., Current topics in cellular regulation. Mechanism and regulation of protein turnover in animal cells, *Curr. Top. Cell. Regul.*, 11, 183, 1976.
6. Rannels, D. E., Kao, R., and Morgan, H. E., Effect of insulin on protein turnover in heart muscle, *J. Biol. Chem.*, 250, 1694, 1975.
7. Khairallah, E. D. and Mortimore, G. E., Assessment of protein turnover in perfused rat liver, *J. Biol. Chem.*, 231, 1375, 1976.
8. Ward, W. F., Cox, J. R., and Mortimore, G. E., Lysosomal sequestration of intracellular protein as a regulatory step in hepatic proteolysis, *J. Biol. Chem.*, 252, 6955, 1977.
9. Poole, B. and Wibo, M., Protein degradation in cultured cells. The effect of fresh medium, fluoride, and iodoacetate on the digestion of cellular protein of rat fibroblasts, *J. Biol. Chem.*, 248, 6221, 1973.
10. Bradley, M. O., Hayflick, L., and Schimke, R. T., Protein degradation in human fibroblasts (WI-38). Effects of aging, viral transformation, and amino acid analogs, *J. Biol. Chem.*, 251, 3521, 1976.
11. Neff, N. T., Ross, P. A., Bartholomew, J. C., and Bissell, M. J., Leucine in cultured cells. Its metabolism and use as a marker for protein turnover, *Exp. Cell Res.*, 106, 175, 1977.
12. Vandenburgh, H. and Kaufman, S., Protein degradation in embryonic skeletal muscle, *J. Biol. Chem.*, 255, 5826, 1980.
13. Gan, J. C. and Jeffay, H., Origins and metabolism of the intracellular amino acid pools in rat liver and muscle, *Biochim. Biophys. Acta*, 148, 448, 1967.
14. Airhart, J., Vidrich, A., and Khairallah, E. A., Compartmentation of free amino acids for protein synthesis in rat liver, *Biochem. J.*, 140, 539, 1974.
15. Vidrich, A., Airhart, J., Bruno, M. K., and Khairallah, E. A., Compartmentation of free amino acids for protein biosynthesis, *Biochem. J.*, 162, 257, 1977.
16. Young, V. R. and Munro, H. N., N^{τ}-Methylhistidine (3-methylhistidine) and muscle protein turnover: an overview, *Fed. Proc. Fed. Am. Soc. Exp. Biol.*, 37, 2291, 1978.
17. Swick, R. W., Measurement of protein turnover in rat liver, *J. Biol. Chem.*, 231, 751, 1958.
18. McFarlane, A. S., Measurement of synthesis rates of liver-produced plasma proteins, *Biochem. J.*, 89, 277, 1963.
19. Swick, R. W. and Ip, M. M., Measurement of protein turnover in rat liver with [^{14}C]carbonate, *J. Biol. Chem.*, 249, 6836, 1974.
20. Scornik, O. A. and Botbol, V., Role of changes in protein degradation in the growth of regenerating livers, *J. Biol. Chem.*, 251, 2891, 1976.
21. MacDonald, M. L., Augustine, S. L., Burk, T. L., and Swick, R. W., A comparision of methods for the measurement of protein turnover *in vivo*, *Biochem. J.*, 184, 473, 1979.
22. Zilversmit, D. B., The design and analysis of isotope experiments, *Am. J. Med.*, 29, 832, 1960.
23. Zilversmit, D. B., Entenman, B. C., and Fishler, M. C., On the calculation of turnover time and "turnover rate" from experiments involving the use of labeling agents, *J. Gen. Physiol.*, 26, 325, 1943.
24. Zak, R., Martin, A. F., and Blough, R., Assessment of protein turnover by use of radioisotopic tracers, *Physiol. Rev.*, 59, 407, 1979.
25. Kay, J., Intracellular protein degradation, *Biochem. Soc. Trans.*, 6, 789, 1978.
26. Poole, B., The kinetics of disappearance of labeled leucine from the free leucine pool of rat liver and its effect on the apparent turnover of catalase and other hepatic proteins, *J. Biol. Chem.*, 246, 6587, 1971.
27. Koch, A. L., The evaluation of the rates of biological processes from tracer kinetic data. I. The influence of labile metabolic pools, *J. Theoret. Biol.*, 3, 283, 1962.
28. Arias, I. M., Doyle, D., and Schimke, R. T., Studies on the synthesis and degradation of proteins of the endoplasmic reticulum of rat liver, *J. Biol. Chem.*, 244, 3303, 1969.
29. Glass, R. D. and Doyle, D., On the measurement of protein turnover in animal cells, *J. Biol. Chem.*, 247, 5234, 1972.

30. **Martin, A. F., Rabinowitz, M., Blough, R., Prior, G., and Zak, R.,** Measurements of half-life of rat cardiac myosin heavy chain with leucyl-tRNA used as precursor pool, *J. Biol. Chem.*, 252, 3422, 1977.
31. **Zak, R. A., Martin, A., Prior, G., and Rabinowitz, M.,** Comparison of turnover of several myofibrillar proteins and critical evaluation of double isotope method, *J. Biol. Chem.*, 252, 3430, 1977.
32. **Dice, J. F., Walker, C. D., Byrne, B., and Cardiel, A.,** General characteristics of protein degradation in diabetes and starvation, *Proc. Natl. Acad. Sci. USA*, 75, 2093, 1978.
33. **Amenta, J. S., Baccino, F. M., and Sargus, M. J.,** Cell protein degradation in cultured rat embryo fibroblasts. Suppression by vinblastine of the enhanced proteolysis by serum-deficient media, *Biochim. Biophys. Acta*, 451, 511, 1976.
34. **Amenta, J. S., Sargus, M. J., and Baccino, F. M.,** Control of cell protein degradation. Changes in activities of lysosomal proteases, *Biochim. Biophys. Acta*, 476, 253, 1977.
35. **Amenta, J. S., Sargus, M. J., Venkatesan, S., and Shinozuka, H.,** Role of the vacuolar apparatus in augmented protein degradation in cultured fibroblasts, *J. Cell. Physiol.*, 94, 77, 1978.
36. **Dell'Orco, R. T. and Guthrie, P. L.,** Altered protein metabolism in arrested populations of aging human diploid fibroblasts, *Mech. Ageing Dev.*, 5, 399, 1976.
37. **Hod, Y. and Hershko, A.,** Relationships of the pool of intracellular valine to protein synthesis and degradation in cultured cells, *J. Biol. Chem.*, 251, 4458, 1976.
38. **Seglen, P. O. and Solheim, A. E.,** Valine uptake and incorporation into protein in isolated rat hepatocytes, *Eur. J. Biochem.*, 85, 15, 1978.
39. **Airhart, J., Kelley, J., Brayden, J. E., and Low, R. B.,** An ultramicro method of amino acid analysis: application to studies of protein metabolism in cultured cells, *Analyt. Biochem.*, 96, 45, 1979.
40. **Wheatley, D. N. and Inglis, M. S.,** An intracellular perfusion system linking pools and protein synthesis, *J. Theoret. Biol.*, 83, 437, 1980.
41. **Johnson, R. W. and Kenney, F. T.,** Intracellular turnover of an altered enzyme, *Intracellular Protein Turnover*, Schimke, R. T. and Katumna, N., Eds., Academic Press, New York, 1975, 163.
42. **Tweto, J. and Doyle, D.,** Turnover of the plasma membrane proteins of hepatoma tissue culture cells, *J. Biol. Chem.*, 251, 872, 1976.
43. **Hayflick, L.,** The limited *in vitro* lifetime of human diploid cell strains, *Exp. Cell Res.*, 37, 614, 1965.
44. **Shakespeare, V. and Buchanan, J. H.,** Increased degradation rates of protein in aging human fibroblasts and in cells treated with an amino acid analog, *Exp. Cell Res.*, 100, 1, 1976.
45. **Dean, R. T. and Riley, P. A.,** The degradation of normal and analogue-containing proteins in MRC-5 fibroblasts, *Biochim. Biophys. Acta*, 539, 230, 1978.
46. **Goldstein, S., Stotland, D., and Cordeiro, R. A. J.,** Decreased proteolysis and increased amino acid efflux in aging human fibroblasts, *Mech. Ageing Dev.*, 5, 221, 1976.
47. **Millward, D. J. and Waterlow, J. C.,** Effect of nutrition on protein turnover in skeletal muscle, *Fed. Proc. Fed. Am. Soc. Exp. Biol.*, 37, 2283, 1978.
48. **Warburton, M. J. and Poole, B.,** Effect of medium composition on protein degradation and DNA synthesis in rat embryo fibroblasts, *Proc. Natl. Acad. Sci. USA*, 71, 2427, 1977.
49. **Ayuso-Parrilla, M. S., Martin-Requero, A., Perez-Diaz, J., and Parrilla, R.,** Role of glucagon on the control of hepatic protein synthesis and degradation in the rat *in vivo*, *J. Biol. Chem.*, 251, 7785, 1976.
50. **Hopgood, M. F., Clark, M. G., and Ballard, F. J.,** Protein degradation in hepatocyte monolayers. Effects of glucagon, adenosine 3′:5′-cyclic monophosphate and insulin, *Biochem. J.*, 186, 71, 1980.
51. **Dazord, A., Gallet, D., and Saez, J. M.,** Protein degradation in adrenal cells in culture is inhibited by ACTH and cyclic AMP, *FEBS Lett.*, 83, 307, 1977.
52. **Baum, B. J., Moss, J., Breul, S. D., Berg, R. A., and Crystal, R. G.,** Effects of cyclic AMP on the intracellular degradation of newly synthesized collagen, *J. Biol. Chem.*, 255, 2843, 1980.
53. **Dannies, P. S. and Rudnick, M. S.,** 2-Bromo-α-ergocryptine causes degradation of prolactin in primary cultures of rat pituitary cells after chronic treatment, *J. Biol. Chem.*, 255, 2776, 1980.
54. **Hershko, A. and Ciechanover, A.,** Mechanisms of intracellular protein breakdown, *Ann. Rev. Biochem.*, 59, 335, 1982.
55. **Amenta, J. S. and Brocher, S. C.,** Mechanisms of protein turnover in cultured cells, *Life Sci.*, 28, 1195, 1981.
56. **Rannels, D. E., Wartell, S. A., and Watkins, C. A.,** The measurement of protein synthesis in biological systems, *Life Sci.*, 30, 1679, 1982.

Chapter 5

APPROACHES TO STUDYING AGE-DEPENDENT CHANGES IN CHROMOSOMAL PROTEINS

Ian Phillips, Anne Shephard, Janet L. Stein, and Gary Stein

TABLE OF CONTENTS

I. Introduction 82

II. Isolation of Nuclei and Chromatin 82
- A. Tissue Culture Cells 83
- B. Mouse and Rat Liver 84

III. Fractionation of Chromosomal Proteins 84
- A. General Considerations 84
- B. Isolation of Histones 86
  - 1. Isolation of the Five Histone Fractions in a Single Procedure 86
  - 2. Molecular Sieve Fractionation of Histones 88
- C. Isolation and Fractionation of Nonhistone Chromosomal Proteins 88
  - 1. Isolation of Nonhistone Chromosomal Proteins 89
    - a. Isolation of Nonhistone Chromosomal Proteins from Dehistonized Chromatin 89
    - b. Isolation of Nonhistone Chromosomal Proteins from Whole Chromatin 89
    - c. Selective Extraction Procedures 89
  - 2. Fractionation of Nonhistone Chromosomal Proteins 93
    - a. Gel Filtration 93
    - b. Ion Exchange Chromatography 93
    - c. Affinity Techniques 93

IV. Polyacrylamide Gel Electrophoretic Techniques for the Analysis of Chromosomal Proteins 95
- A. General Information 96
  - 1. Washing of Gel Plates and Tubes 96
  - 2. Materials 96
- B. Analysis of Total Chromosomal Proteins or Nonhistone Chromosoma Proteins 96
  - 1. One-Dimensional Gel Electrophoretic Systems 96
    - a. SDS-Polyacrylamide Gel Electrophoresis 96
    - b. Isoelectric Focusing Gel Electrophoresis 98
  - 2. Two-Dimensional Gel Electrophoretic Systems 98
    - a. Isoelectric Focusing — SDS-Polyacrylamide Gel Electrophoresis 98
    - b. Nonequilibrium pH Gradient Gel Electrophoresis (NEPHGE)/SDS-Polyacrylamide Gel Electrophoresis 101
- C. Analysis of Histones 101
  - 1. One-Dimensional Gel Electrophoretic Systems 101
    - a. Acid-Urea Polyacrylamide Gel Electrophoresis 101
    - b. Triton-Acid-Urea Gel Electrophoresis 102
    - c. SDS-Polyacrylamide Gel Electrophoresis 102

2. Two-Dimensional Gel Electrophoretic Systems 104
a. Acid-Urea/SDS-Polyacrylamide Gel Electrophoresis ..104
b. Acid-Urea/Triton-Acid-Urea 104
c. NEPHGE/SDS-Polyacrylamide Gel Electrophoresis ..104
D. Analysis of Radiolabeled Proteins 104
1. Gel Fractionation and Counting..........................105
2. Fluorography 105
3. Autoradiography 105

Acknowledgments 106

References...106

I. INTRODUCTION

Cellular aging in intact organisms as well as in vitro is associated with a broad spectrum of biochemical modifications. Such age-dependent biochemical changes at least in part reflect alterations in gene expression — undoubtedly at several levels within the nucleus and cytoplasm. Chromosomal proteins, histones and nonhistone chromosomal proteins, have been shown to play a central role in the structural and the transcriptional properties of the eukaryotic genome. Hence examining age-dependent alterations in the composition and metabolism of chromosomal proteins as well as interactions of chromosomal proteins with defined genetic sequences as a function of age may provide additional insight into mechanisms operative in the aging process at the cellular and molecular levels.

In this chapter we will attempt to present methodologies which have proven to be effective for studying genome-associated proteins. Whenever possible we will consider procedures appropriate for examining chromosomal proteins from cells in culture and from tissues of intact organisms in a parallel manner. In addition to presenting recipes, we will evaluate the strengths and weaknesses of each procedure.

II. ISOLATION OF NUCLEI AND CHROMATIN

One of the primary considerations in approaching the analysis of chromosomal proteins is the selection of methods for isolation of nuclei and chromatin. We cannot overemphasize that the procedures employed have a very direct bearing on the final product. In fact, nuclei and chromatin could be appropriately defined operationally by the isolation procedures. It therefore follows that the validity of chromatin as an accurate representation of the genome found in the nucleus of intact cells, and hence the representation of genome-associated proteins, is directly dependent on the methods employed for isolation of nuclei and chromatin.

Although a broad spectrum of approaches has been utilized for isolation of nuclei and chromatin, there is no single procedure that appears to be optimal for all organisms, tissues, and cell types. Rather, modifications in protocols are necessary to accommodate specific biological situations. The following are general considerations which should determine the acceptability of a technique: (1) It is generally best to prepare nuclei free of cytoplasmic material and stripped of the outer component of the nuclear envelope. This can best be achieved by treatment with citric acid buffer or with non-ionic detergents such as Triton® X-100 or NP-40. The concentration of detergent which can be tolerated varies depending on the cell type. (2) An effort should be made to eliminate material present in nuclear sap. (3) Caution must be exercised to avoid extraction of proteins bound to chromatin. (4) An attempt should be made to shear DNA as little as possible. (5) Nuclease and protease activity should be minimized.

We will concentrate on procedures which we use in our laboratories for isolation of nuclei and chromatin from two lines of tissue culture cells, HeLa S_3 cells (human cervical carcinoma cells grown in suspension culture) and WI-38 human diploid fibroblasts at various passages (embryonic lung cells grown in monolayer culture), as well as the procedure we utilized for the preparation of nuclei and chromatin from mouse and rat liver. Alternative protocols for nuclei and chromatin isolation will also be considered.

A. Tissue Culture Cells

All procedures are carried out at 4°. All procedures can be carried out in the tube in which cells are initially harvested, resulting in 85 to 95% recovery of chromatin. Avoiding sonication or homogenization steps results in minimal shearing of the DNA. We have observed that shearing can result in a significant increase in chromatin template activity with exogenous RNA polymerase. Cells grown in suspension culture are harvested by centrifugation at 1000 g for 5 min, and cells grown in monolayers are scraped from the culture vessel with a rubber policeman and then collected by centrifugation at 1000 g for 5 min. Treatment with trypsin should be avoided since the enzyme can utilize chromosomal proteins as substrates. The harvested cells are washed three times with Spinner Salts (Grand Island Biological Co., Grand Island, New York) to remove serum proteins — each wash step followed by centrifugation at 1000 g for 5 min. Cells are lysed by resuspension in 80 volumes of 80 m*M* NaCl — 20 m*M* EDTA — 1% Triton® X-100 (pH 7.2) and agitated with a vortex mixer for 20 sec at maximum speed. Nuclei are pelleted by centrifugation at 1000 g for 5 min in a swinging bucket rotor and then washed twice with 80 m*M* NaCl-20 m*M* EDTA-1% Triton® X-100. The nuclei should now be free of visible cytoplasmic material when examined by phase contrast microscopy — this should be a routine procedure. Electron microscopic examination of nuclei should reveal the absence of the outer aspects of the nuclear envelope and often the inner component of the nuclear envelope is also removed by detergent treatment. Nuclei are washed twice with 80 volumes of 0.15 *M* NaCl-10 m*M* Tris (pH 8.0) and centrifuged at 1000 g for 5 min to remove the detergent. It is important to remove the supernatant completely following the last wash step since the salt will interfere with nuclear lysis. Nuclei are lysed by resuspension in double distilled water (at a concentration of 250 to 500 μg DNA/mℓ) by gentle agitation with a vortex mixer. Nuclear lysis results in a marked increase in viscosity after swelling in an ice bath for 20 min. The material is clear and gelatinous. Incomplete removal of cytoplasm during isolation, incomplete lysis, and protein denaturation will be reflected by a "cloudy" or "milky" appearance of the chromatin gel. Chromatin is pelleted by centrifugation at 12,000 g for 20 min in a fixed angle rotor, and the chromatin is again washed in distilled water and pelleted at 12,000 g. The chromatin pellet should be clear to slightly opalescent and extremely gelatinous. Chromatin prepared by this method has a protein:DNA ratio of 1.8 to 2.0. The histone: DNA ratio is 1:1 and the nonhistone chromosomal protein:DNA ratio is 0.9.

Depending upon the cell line and specific procedure employed, varying extents of protease activity may be encountered during preparation of nuclei and chromatin. In some instances it may therefore be advisable to include protease inhibitors in buffers. Some of the inhibitors which have been reported to be effective at preventing proteolysis of nuclear and chromosomal proteins include sodium bisulfite, L-1-tosylamide-2-phenyl-ethyl-chloromethyl-ketone, phenylmethylsulfonyl fluoride, and diisopropyl fluorophosphate. Another approach to nuclear isolation which has been shown to be effective in minimizing proteolytic activity and at the same time permits maximal retention of nuclear and chromosomal proteins (enzymes and structural proteins) employs nonaqueous media. Nonaqueous procedures for cell fractionation were introduced several decades ago by Behrens[1,2] and modified for nuclear isolation by Allfrey et al.[3] Details of recently optimized nonaqueous nuclear isolation protocols are contained in articles by Gurney and Foster[4] and Wray et al.[5]

B. Mouse and Rat Liver

Animals are killed by cervical dislocation. All lobes of the liver are excised, placed in a plastic weighing boat on ice, and weighed. Nuclei are prepared by modification of the method of Cheveau et al.[6] The liver is immediately minced with surgical scissors and suspended in 20 volumes of 2.2 *M* sucrose — 4 m*M* magnesium chloride. All procedures are carried out at 4° centigrade. The liver is homogenized to homogeneity in a motor driven Potter Elvehjem homogenizer with a wide clearance Teflon® pestle. The homogenate is filtered through one layer of Miracloth (Chicoppee Mills, New York) and centrifuged in a Beckman SW27 rotor for 60 min at 25,000 rpm. The supernatant is discarded and after removal of material adhering to the walls of the centrifuge tube, the nuclear pellet is suspended in 80 volumes of 0.15 *M* NaCl-10 m*M* Tris (pH 8.0). After centrifugation at 15,000 g in a swinging bucket rotor for 3 min, the nuclei are again washed in 0.15 *M* NaCl-10 m*M* Tris (pH 8.0). The latter two washing steps deplete the nuclei of sucrose. To remove the outer aspect of the nuclear envelope and adhering cytoplasmic material, nuclei are resuspended in 80 volumes of 80 m*M* NaCl-20 m*M* EDTA-1% Triton® X-100 (pH 7.2) and agitated with a vortex mixer for 10 sec at maximal speed. Nuclei are pelleted by centrifugation at 1000 g for 5 min in a swinging bucket rotor and then washed with 80 volumes of 0.15 *M* NaCl-10 m*M* Tris (pH 8.0) and centrifuged at 1000 g for 5 min to remove the detergent. We have observed that liver from certain strains of mouse or rat requires lower concentrations of detergent (as low as 0.1%) to avoid nuclear lysis during detergent treatment. Nuclear lysis and recovery of chromatin are carried out as described above for tissue culture cells.

The same precautions employed for minimizing proteolytic degradation during isolation of nuclei and chromatin from tissue culture cells should be applied to nuclear and chromatin isolation from tissues of intact animals. It should also be noted that nuclear isolation from tissue in the presence of nonaqueous reagents is highly effective.

III. FRACTIONATION OF CHROMOSOMAL PROTEINS

A. General Considerations

Fractionation of chromosomal proteins represents a formidable task, yet one which must be met if the structural and functional properties of chromosomal polypeptides are to be elucidated. The nonhistone chromosomal proteins are extremely heterogeneous, minimally 500 classes of polypeptides, and recent observations suggest histone microheterogeneity, rendering these proteins more complex than the five species previously believed to exist. Additionally, major components of nonhistone chromosomal proteins are insoluble under conditions employed for routine protein fractionation. Yet in spite of these obstacles some progress has been made towards separating and characterizing chromosomal proteins.

In evaluating procedures for fractionation of chromosomal proteins, the same considerations which apply for isolation of nuclei and chromatin are operative, e.g., absence of cytoplasmic material and minimization of proteases and nucleases are imperative. Since fractionation of chromosomal proteins frequently requires numerous steps and extensive periods of time, precautions to avoid degradation and irreversible denaturation are particularly important. It should be emphasized that approaches to chromosomal protein fractionation must often be modified to accommodate differences inherent in various tissues and cell types.

It is necessary to consider both the loss of genome-associated proteins and the adherence of cytoplasmic, membrane, and nucleoplasmic components during chromatin preparation. Chromatin extracted from purified nuclei[7] contains less cytoplasmic contamination than chromatin isolated directly from whole tissues.[8] Cytoplasmic and membrane contamination of chromatin can be reduced by washing nuclei in solutions containing nonionic detergents such as Triton® X-100 (described above), or by centrifuging the chromatin through heavy sucrose. When chromatin is prepared by these techniques, in the presence of radioactively labeled cytoplasmic extracts, less than 5% of HeLa cell chromatin proteins are due to cytoplasmic contaminants.[9,10] The use of cytoplasmic "marker" enzymes reveals less than 10% cytoplasmic contamination of chromatin.[11] When considering this problem, it should be remembered that proteins can migrate in vivo from the cytoplasm to the chromatin; e.g., chromosomal proteins are synthesized in the cytoplasm and are then transported into the nucleus,[12] and hormonal stimulation[13] or cell fusion[14] causes the migration of some cytoplasmic proteins into the nucleus. During the isolation of nuclei in aqueous media, proteins can be lost from or absorbed into the nucleus. As discussed previously, this problem can be overcome by isolating nuclei in nonaqueous media. The presence of proteins common to both nucleoplasm and chromatin fractions need not necessarily be interpreted as an artifact of preparation. Rather, this similarity may reflect a nucleoplasmic pool of macromolecules which exists in a dynamic equilibrium with the chromatin.[15] This movement of macromolecules between the cytoplasm, nucleoplasm, and chromatin most likely has functional importance.

A problem sometimes encountered in fractionating nonhistone chromosomal proteins is that fractions may contain nucleic acid — RNA as well as DNA. Such nucleic acids can complicate experiments in which nonhistone chromosomal protein fractions are assayed in chromatin reconstitution systems for ability to render specific genes transcribable. If nucleic acids are covalently bound to the protein fractions, separation of nucleic acids and chromosomal proteins is extremely difficult. However, separation of noncovalently associated nucleic acids and chromosomal proteins can be executed by buoyant density centrifugation in solutions of cesium salts and urea.

Buoyant density centrifugation in cesium chloride has been used by Gilmore and Paul[16] to prepare chromosomal proteins under conditions that exclude endogenous RNA. Chromatin is dissolved in 55% cesium chloride-4 *M* urea-10 m*M* Tris-HCl (pH 8.0)-10 m*M* dithiothreitol-10 m*M* EDTA at a concentration of 300 μg/mℓ. Four 5-mℓ aliquots of dissociated chromatin are then centrifuged at 40,000 rpms for 40 hr at 8°C in an MSE 10 × 10 titanium rotor or the equivalent (the remainder of the tube is filled with paraffin oil). The chromosomal proteins are found in the top 1.5 mℓ of the gradient, while the DNA and RNA form a pellet. A procedure which has been used successfully by Modak and co-workers[17,18] to separate DNA, RNA, and protein involves centrifugation in cesium sulfate and urea. This method is particularly useful when the isolation of the nucleic acid components is desired. The sample is adjusted to 5 *M* urea— 0.54 mg/mℓ cesium sulfate-30 m*M* NaCl-20 m*M* Tris-HCl (pH 7.4)-5 m*M* EDTA. 15 mℓ aliquots, overlaid with paraffin oil, are centrifuged for 50 to 62 hr at 170,000 g in a Beckman 60 Ti rotor at 22°. RNA forms an unprecipitated band at density 1.7 g/cm^3, DNA bands at 1.5 g/cm^3 and protein floats.

B. Isolation of Histones

A number of procedures have been developed for the preparative scale isolation of highly purified histone fractions. Among the primary considerations in selection of a specific protocol should be the cells or tissues from which the histones are to be isolated, the quantity of the histones required, the extent of denaturation tolerable, whether a single histone fraction is required or whether it would be desirable to isolate all histone fractions during one preparation, and the extent to which the histone fractions are irreversibly modified during isolation. Purified histone fractions may be prepared by selective extraction with salts, ionic detergents, or dilute mineral acids; precipitation with various concentrations of trichloroacetic acid; ion exchange chromatography and molecular sieving. While it would be unrealistic to present detailed protocols for all of the available histone preparations, we will present several which have proven to be highly reproducible and represent the different approaches to isolation of these chromosomal polypeptides. Other equally effective procedures are detailed in *Methods in Cell Biology*.[19-22]

1. Isolation of the Five Histone Fractions in a Single Procedure

For more than a decade Johns and co-workers have pioneered the development and refinement of histone isolation procedures. The specific technique detailed below for isolation of the five principal histone fractions and further described elsewhere[23] represents an amalgamation of earlier procedures developed by Johns for preparation of histones.[24-26]

Unless otherwise indicated, all procedures are carried out at 4°C. The procedure as described is for preparation of histones from 50 g of tissue or cultured cells. With appropriate adjustments larger or smaller amounts of material may be used. The yield from a 50-g preparation is approximately 200 to 300 mg of each histone fraction.

- Step 1: *Animal tissue:* Immediately after killing the animal, preferably by cervical dislocation, the tissue is excised, connective tissue is removed and 50 g of material are minced with sharp surgical scissors or a single-edge razor blade. If histone isolation is not to proceed immediately, the tissue is frozen in liquid nitrogen and stored at −80°C. *Cultured cells:* Suspension cells are harvested by centrifugation and monolayer cells are scraped with a rubber policeman and pelleted by centrifugation as described in a previous section of this chapter. The cells are washed at least twice with Earls balanced salt solution (Grand Island Biological Co., Grand Island, New York). All subsequent steps are identical for cells derived from tissue or cultures.
- Step 2: Homogenize in a Waring® blender for 2 min at maximum speed in 700 mℓ of 0.14 *M* NaCl adjusted to pH 5.0 with HCl to minimize proteolytic degradation. Adjust the pH of the homogenate to 5.0 by addition of HCl. Centrifuge the homogenate at 1500 g for 30 min. Discard supernatant.
- Step 3: Resuspend pellet in 700 mℓ of 0.14 *M* NaCl (pH 5.0) and homogenize in blender for 1 min. Centrifuge at 1500 g for 20 min. Discard supernatant. Repeat Step 3 four times, homogenizing for 30 sec and centrifuge for 15 min.
- Step 4: The pellet is suspended in 700 mℓ of 90% ethanol, blended at half speed for 30 sec and centrifuged at 1500 g for 15 min. The supernatant is discarded.
- Step 5: The pellet is suspended in 400 mℓ of ethanol-1.25 *N* HCl (4:1, v/v), transfered to a 1-ℓ wide mouth polyethylene bottle and 400 mℓ of mixed glass beads (4 to 8 mm in diameter) are added. The material is agitated in a gyratory or reciprocating shaker overnight (12 to 18 hr depending on the sleep requirements of the investigator). The suspension is filtered through nylon screening to

eliminate the glass beads and the deoxynucleoprotein slurry is centrifuged at 1500 g for 15 min. The supernatant which contains histones H3, H2A, and H4 is saved.

- Step 6: The pellet is extracted twice more as described in Step 5 using 200 mℓ of ethanol-1.25 *N* HCl (4:1) and extracting 3 hr each time. The supernatants from Steps 5 and 6 are combined (approximately 800 mℓ).
- Step 7: The combined supernatants from Steps 5 and 6 are clarified by filtration through a number 4 sintered glass funnel and dialyzed twice against 4.5 volumes of anhydrous ethanol overnight at 22°C.
- Step 8: The precipitate which forms during dialysis (H3 histone) is collected by centrifugation, washed twice in 100 mℓ of the solution against which it had been dialysed, washed 3 times in acetone and then dried under vacuum or under a gentle stream of nitrogen. If the pellet is broken up with a glass rod during the drying procedure, a fine white powder will result which is completely water soluble.
- Step 9: The supernatant which remains after precipitation of H3 histone, containing H4 and H2A histone, is dialyzed for 4 hr against the same solution after addition of 2 volumes of ethanol. The dialyzed proteins are placed at −10° for 12 to 16 hr and then brought to 22°. A small amount of precipitation occurs and this material is removed by centrifugation and discarded. The supernatant is filtered through a number 4 sintered glass funnel.
- Step 10: 44 g of solid guanidine hydrochloride and 8.6 mℓ of HCl are added to the filtered supernatant while rapidly stirring until the guanidine hydrochloride is dissolved. The solution is then clarified by filtering through a number 4 sintered glass funnel. At this point in the procedures the volume of the H2A/H4 histone-containing solution should be approximately 660 mℓ 1.25 volumes of acetone are added during stirring and the solution is stirred for approximately 15 min during which time H2A histone precipitates. The precipitate is pelleted by centrifugation, washed twice in a solution similar to that used for precipitation of H2A, washed three times in acetone and then dried under a vacuum or in a gentle stream of nitrogen as described in Step 8.
- Step 11: The supernatant containing H4 histone is clarified by filtration through a number 4 sintered glass funnel and an equal volume of acetone is added. The H4 histone precipitate is pelleted by centrifugation, washed 3 times in acetone and dried under a vacuum or under nitrogen as described in Step 8. Isolation of H1 and H2B histones from the sediment which remains following extraction of H2A, H3, and H4 histones from saline washed nucleoproteins (Steps 1—6) proceeds as follows.
- Step 12: The pellet is extracted twice with 200 mℓ of ethanol: 1.25 *N* HCl (4:1, v/v) for 1.5 hr each as described in Steps 5 and 6, but now the supernatant following centrifugation is discarded.
- Step 13: The pellet is similarly extracted with 200 mℓ of 0.25 *N* HCl for 18 hr followed by two extractions with 100 mℓ of 0.25 *N* HCl for 2 hr each. The combined supernatants containing H1 and H2B histones are clarified by filtration through a number 4 sintered glass funnel, brought to 22°C, and 3 volumes of acetone are added to precipitate H1 histone. The precipitate is washed in 100 mℓ of 0.25 *N* HCl:acetone (1:3, v/v) at room temperature, then 3 times in acetone, and dried under a vacuum or nitrogen as described above.
- Step 14: The supernatant containing H2B histone is filtered through a sintered glass funnel and precipitated by addition of two volumes of acetone. The precipitated H2B histone is collected by centrifugation, washed 3 times in acetone, and dried under vacuum or nitrogen.

2. Molecular Sieve Fractionation of Histones

Von Holt and co-workers have developed several highly effective procedures for fractionation and purification of individual histone fractions[27-31] exploiting the ability of these chromosomal polypeptides to separate on molecular sieve matrices. (A detailed description of these procedures is found elsewhere.[32]) This approach obviously takes advantage of molecular weight variations between the various histone fractions (although histones migrate anomolously with respect to their actual size on molecular sieve matrices) and in addition, depends in part on noncovalent interactions between histone arginine residues and carboxyl groups of BioGel®, conformational changes brought about by solutes and solvents, histone-histone interactions and the reversible dimerization of H3 histones. While there are a variety of approaches which can be taken to exploit molecular sieves for histone fractionation (see Von Holt and Brant for review and details[32]), in this chapter we will present a simple two-step procedure by these investigators which permits reproducible preparation of histone fractions in a minimal period of time. It should be emphasized that we are using calf thymus and chicken erythrocyte histones as an example and that modification in the protocol is required depending upon the species and tissue from which the histones are derived.

- Step 1: The histone sample, extracted from purified nuclei or chromatin by dilute mineral acid extraction, direct salt extraction or dissociation of histones from DNA with protamine, is dissolved in 8 *M* urea-1% β mercaptoethanol-0.02 *N* HCl.
- Step 2: The samples are applied to a 90 cm Biogel® P-60 column (15 to 20 mg/cm^2 column cross-section area) equilibrated with 0.02 *N* HCl-0.02 *N* NaN_3 (pH 1.7)-0.05 *M* NaCl (20-25°) and eluted with the same buffer. In the case of calf thymus, Hl, H2B, and H4 histones elute as individual peaks while H3 and H2A histones co-fractionate. In the case of chicken erythrocytes, H1, H5, H2A, and H4 histones elute as individual peaks while H2B and H3 co-fractionate.
- Step 3: Fractions are dialyzed against distilled water at 4°, lyophilized to dryness and stored at −80°.
- Step 4: The H2A-H3 calf thymus histone fraction and the H2B-H3 chicken erythrocyte histone fraction are solubilized in 0.05 *M* sodium acetate-0.005 *M* sodium bisulfite (pH 5.1) at 20 to 25°C.
- Step 5: The histone samples are applied to a 90 cm Sephadex® G-100 column (15 to 20 mg/cm^2 column cross-section area) and eluted with 0.05 *M* sodium acetate-0.005 *M* sodium bisulfite (pH 5.1) at 20 to 25°.
- Step 6: The histone fractions are dialyzed against distilled water at 4° and lyophilized to dryness.

C. Isolation and Fractionation of Nonhistone Chromosomal Proteins

Ideally, a procedure for isolation and fractionation of nonhistone chromosomal proteins should permit high yields, recovery of all protein species, separation of nonhistone chromosomal proteins from nucleic acids (DNA and RNA) and histones, avoidance of harsh denaturing conditions which may alter the structural and functional properties of the proteins, and fractionation of these heterogeneous chromosomal polypeptides into subfractions. However, in approaching the isolation and fractionation of nonhistone chromosomal proteins, it is necessary to keep in mind that difficulties are encountered because of the complexity of the nonhistone chromosomal proteins, the insolubility of certain nonhistone chromosomal protein components, and the tendency of nonhistone chromosomal protein species to form aggregates with other nonhistone chromosomal proteins as well as with histones or nucleic acids. What has emerged is a series of approaches which fall short of the criteria set forth above but which do, indeed, permit assessment of the representation of the nonhistone chromosomal polypeptides under a variety of biological conditions and evaluation of the structural and functional properties of nonhistone chromosomal fractions.

1. Isolation of Nonhistone Chromosomal Proteins

Methods for the isolation of nonhistone chromosomal proteins fall into three general categories: (1) those involving selective extraction of histones followed by dissociation and separation of nonhistones from nucleic acid; (2) those involving the dissociation of macromolecules from total chromatin and their subsequent fractionation; and (3) selective extraction of nonhistone chromosomal proteins directly from chromatin.

a. Isolation of Nonhistone Chromosomal Proteins from Dehistonized Chromatin

Histones are first extracted from chromatin, usually with dilute HCl or H_2SO_4; then the nonhistone chromosomal proteins are separated from nucleic acids by various techniques. In early experiments in which the objective was merely to determine the quantity, chemical composition, or radioisotope content of the nonhistone chromosomal proteins, nucleic acids where hydrolyzed with 5% trichloroacetic acid or 0.5 *N* perchloric acid at 90° and the nonhistone chromosomal proteins solubilized in strong alkali.[33-36] This method completely denatures the proteins, and it makes them unsuitable for further investigation. To enable the nonhistone chromosomal proteins to be analyzed by techniques such as gel electrophoresis, less drastic methods of extraction were developed (Table 1). It should be noted that "dehistonization" of chromatin by dilute mineral acid extraction may alter the nonhistone proteins such that their extractability and biological properties may be irreversibly modified. Therefore, there is considerable advantage to dehistonization by milder conditions such as urea-sodium chloride at pH 6.0.[42]

b. Isolation of Nonhistone Chromosomal Proteins from Whole Chromatin

In procedures of this type the chromatin is dissociated and solubilized; then the DNA and histones are separated from the nonhistone chromosomal proteins (Table 2). Methods that involve the coextraction of histones and nonhistone chromosomal proteins result in a reduced yield of nonhistone proteins because (1) before histones can be removed by ion exchange chromatography the salt concentration of the sample must be reduced to less than 0.5 *M*, usually by dialysis, which results in considerable losses of the nonhistone chromosomal proteins due to precipitation[63,65] and (2) approximately 10% of the nonhistone proteins chromatograph with the histones.[63,65,69]

c. Selective Extraction Procedures

Methods discussed in this section have been used to selectively extract fractions of the nonhistone chromosomal proteins. Gronow and co-workers extracted 70% of the total nuclear proteins with 8 *M* urea-50 m*M* phosphate (pH 7.6)[77-79] and other groups have used this technique for selectively extracting a fraction of nonhistone chromosomal proteins from chromatin.[80] A group of nonhistone chromosomal proteins enriched in phosphoproteins was extracted by Langen[80a] using a method similar to that of Wang.[57] Modifications to this method were introduced by Gershey and Kleinsmith[81] and Kleinsmith and Allfrey.[81a] Histones were removed using Biorex 70 and phosphoproteins were adsorbed to calcium phosphate gel. This phosphoprotein enriched fraction comprised less than 30% of the total nonhistone chromosomal proteins, and because of the low solubility of these proteins in dilute buffer, they were usually solubilized in SDS.[82,83]

Treatment of chromatin with 0.35 *M* NaCl was first introduced as a method of removing contaminating cytoplasmic proteins from chromatin.[8] Later it became evident that the proteins extracted by this procedure were of chromosomal origin and

Table 1
EXTRACTION OF NONHISTONE CHROMOSOMAL PROTEINS FROM DEHISTONIZED CHROMATIN

Solubilization of dehistonized chromatin	Removal of nucleic acid	Comments	Ref.
4 *M* CsCl (pH 11.6)	Equilibrium density gradient centrifugation	40% yield	37
4 *M* CsCl (pH 14.0)	Equilibrium density gradient centrifugation	Increased yield	38
0.1 to 1% SDS	DNA pelleted by ultracentrifugation	90 to 95% yields. All the major nonhistone chromosomal proteins were recovered; detergent causes denaturation and is difficult to remove	39
			40
1% SDS, 8 M urea (pH 11.5), 5 m*M* 2-mercaptoethanol			41
Histones (together with 15% of the nonhistones) revoved with 2 *M* NaCl, 5 *M* urea (pH 6). Nonhistones extracted from residue by increasing pH to 8.5			42
	DNase I treatment	Yield 95%; nonhistone chromosomal proteins were then precipitated in 0.4 *M* $HClO_4$ and dissolved in SDS	43
		Dissolved precipitate in 10 *M* urea, 0.9 *M* acetic acid, 1% 2-mercaptoethanol	44
0.1 *M* Tris-HCl (pH 8.4), 0.01 *M* EDTA, 0.14 *M* 2-mercaptoethanol	Nonhistone chromosomal proteins partitioned into phenol	Some of residual protein solubilized by re-extracting with phenol at pH 9.5. Nonhistone chromosomal proteins reconstituted into aqueous media by dialyses.	45—48
0.1 *M* Tris-HCl (pH 8.4), 0.01 *M* EDTA, 0.14 *M* 2-mercaptoethanol	Nonhistone chromosomal proteins partitioned into phenol	Phenol-soluble proteins precipitated and then dissolved in 3% SDS, 1% 2-mercaptoethanol	49—51
0.1 *M* Tris-HCl (pH 8.4), 0.01 *M* EDTA, 0.14 *M* 2-mercaptoethanol	Nonhistone chromosomal proteins partitioned into phenol	Phenol extracts dialyzed directly against electrophoresis sample buffer. Residual protein extracted with 5% SDS, 0.14 *M* 2-mercaptoethanol at 100°.	52, 53
0.1 *M* Tris-HCl (pH 8.4), 0.01 *M* EDTA, 0.14 *M* 2-mercaptoethanol	Phenol saturated with 0.1 *M* H_2SO_4		54

several groups have since used this method to isolate a group of relatively loosely bound nonhistone chromosomal proteins[84-89] which may be important in the maintenance of chromatin structure and in the regulation of gene expression. The salt-urea extraction method has been adapted to extract several different fractions of nonhistone chromosomal proteins by the sequential treatment of chromatin with a series of 5 *M* urea solutions containing progressively higher concentrations of NaCl.[90] Nonhistone chromosomal proteins have also been selectively extracted from chromatin by the use of alkaline reagents.[91,92] The proteins extracted were significantly different from those which remained bound to the chromatin and their molecular weight and metabolism were higher.

Table 2
ISOLATION OF NONHISTONE CHROMOSOMAL PROTEINS FROM WHOLE CHROMATIN

Dissociation of chromatin	Removal of DNA	Removal of histones	Comments	Ref.
1 to 3 *M* NaCl	Reduced salt concentration to 0.14 *M*		Leaves majority of nonhistone chromosomal proteins in solution	56, 57
1 to 3 *M* NaCl	Biogel® A-50 m	Urea/polyacrylamide gels (pH 2.7)	Recovery 30 to 60%	58
		Cation exchange resins		59, 60
Guanidine hydrochloride	Gel filtration with Sepharose 4B	Adsorption to CM-Sephadex	Yield 60 to 70%	61
Guanidine hydrochloride	Biogel® A-5m	Adsorption to CM-Sephadex	Yield 60 to 70% *Disadvantage:* gel filtration methods result in cross-contamination of DNA and protein, and dilution of protein	62
Guanidine-HCl	Ultracentrifugation	Ion-exchange chromatography		63
Guanidine-HCl + urea	Ultracentrifugation	Ion-exchange chromatography	90% yield	64, 65
2 to 3 *M* NaCl containing 5 to 7 *M* urea	Pelleted by ultracentrifugation	Ion-exchange chromatography	Introduction of urea into the salt solution enabled chromatin to be dissociated without extensive shearing.	58, 66
			>90% recovery	67, 68
2 to 3 *M* NaCl containing 5 to 7 *M* urea	Pelleted by ultracentrifugation	Nonhistone chromosomal proteins adsorbed to anion exchange resins		69, 70
2 to 3 *M* NaCl containing 5 to 7 *M* urea	Precipitation with $LaCl_3$ (0.0135 *M*)	Ion-exchange chromatography		71

Table 2 (continued)
ISOLATION OF NONHISTONE CHROMOSOMAL PROTEINS FROM WHOLE CHROMATIN

Dissociation of chromatin	Removal of DNA	Removal of histones	Comments	Ref.
2 to 3 *M* NaCl containing 5 to 7 *M* urea	Column chromatography on hydroxylapatite using a phosphate buffer gradient		60 to 70% recovery	72, 73
2 to 3 *M* NaCl containing 5 to 7 *M* urea	Column chromatography on hydroxylapatite using a guanidine hydrochloride gradient		Almost 100% recovery	74
1% SDS	Ultracentrifugation	Ion-exchange chromatography	Difficult to remove histone from SDS solutions	75
25% formic acid, 0.2 *M* NaCl, 8 *M* urea	Ultracentrifugation	Ion-exchange chromatography		76

2. Fractionation of Nonhistone Chromosomal Proteins

a. Gel Filtration

Fractionation of nonhistone chromosomal proteins by gel filtration can be carried out on columns of Sepharose,[61] Sephadex,[67,76,93,94] agarose,[58] and polyacrylamide Biogels.[67] To circumvent the low solubility of nonhistone chromosomal protein components, molecular weight sieving should be carried out in the presence of high salt and/or with 5-7 *M* urea.

b. Ion Exchange Chromatography

Wang and co-workers[95-97] were able to achieve a limited extent of nonhistone chromosomal protein fractionation using DEAE cellulose and low salt conditions. Levy et al.[64] and Chaudhuri[69] were able to overcome the aggregation problem by fractionating nonhistone chromosomal proteins on DEAE cellulose in the presence of 3-5 *M* urea. Using a linear NaCl gradient Levy et al.[64] found two major peaks of proteins from rabbit chromatin and with a more complex gradient these were separated into additional distinct peaks. Several workers[68,69,98-101] have used QAE Sephadex® with a NaCl gradient to fractionate nonhistone chromosomal proteins in the presence of 5 *M* urea. Up to 70% of the protein was removed from the column but Augenlicht and Baserga[65] obtained 85 to 90% recovery when they used a final elution step involving 4 *M* guanidine hydrochloride-6 *M* urea and Rickwood and MacGillivray[74] obtained almost complete recovery by using guanidine hydrochloride instead of sodium chloride for a gradient elution. Elgin and Bonner[76] recovered 4 fractions of nonhistone chromosomal proteins using the cation exchange resin SE-Sephadex® C-25 at an acidic pH. CM Sephadex® has also been used to fractionate the nonhistone chromosomal proteins.[61,102]

Stein et al.[103] have been able to fractionate HeLa cell nonhistone chromosomal proteins on QAE Sephadex® as follows. Chromosomal proteins which have been dissociated from chromatin by high concentrations of salt and urea and from which nucleic acid has been removed by ultracentrifugation are dialyzed against 4 changes of 10 volumes of 5 *M* urea-10 m*M* Tris HCl (pH 8.3) at 4°. The proteins are loaded on a column of QAE Sepahdex® G-25 which has been equilibriated with this same buffer at a flow rate of 15 to 30 mℓ/cm²/hr using approximately 1 g of Sephadex® per 10 mg of proteins. The column is eluted with 2.5 column volumes each of 5 *M* urea-10 m*M* Tris-HCl (pH 8.3) containing 0 *M*, 0.1 *M*, 0.25 *M*, 0.5 *M* or 3 *M* NaCl at the same flow rate (Figure 1) The total recovery of proteins is 85%. The SDS-polyacrylamide gel electrophoretic profiles of a column fraction are shown in Figure 2.

c. Affinity Techniques

The ability of some nonhistone chromosomal proteins to bind to DNA has been exploited as a method of fractionating proteins. Teng et al.[48] extracted nonhistone proteins in high salt solutions and allowed them to bind to DNA during gradient dialysis to reduce the salt concentration to 0.01 *M*. One of the problems associated with this procedure is that the low salt concentrations cause some of the proteins to aggregate and precipitate. Some workers have overcome this dilemma by performing the binding reaction in the presence of 5 *M* urea.[104] The DNA-protein complexes are recovered by sucrose density gradient centrifugation.[47,48,105] DNA binding proteins have also been isolated by affinity chromatography on DNA bound to cellulose,[60,106,107] agarose, polyacrylamide,[108] or Sepharose. However in the low salt concentrations used in these studies many proteins bind to DNA in a nonspecific manner. This can be avoided either by carrying out the binding reaction in higher salt concentration such as 0.2 to 0.25 *M*[109] or by prerunning the proteins through a column of heterologous DNA.[60,107] Allfrey and co-workers[110-112] bind single or double stranded DNA of differ-

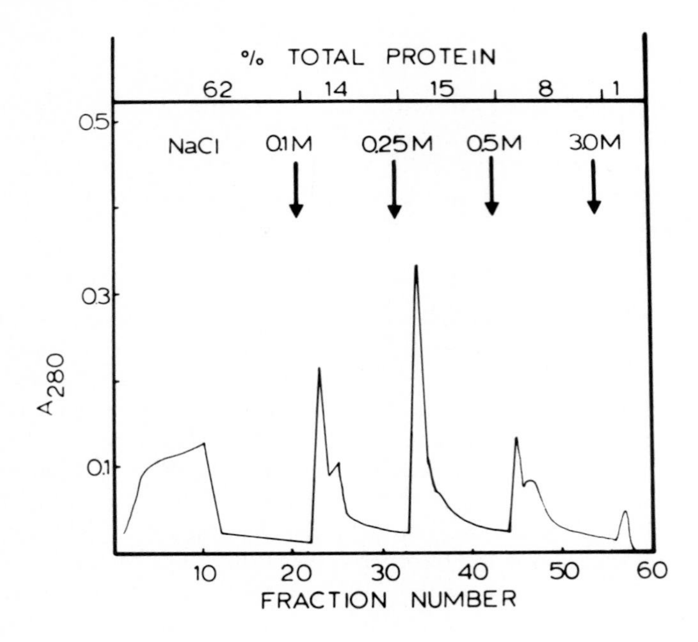

FIGURE 1. Elution profile of chromosomal proteins from QAE-Sephadex®. Proteins were loaded in 5 *M* urea-10 m *M* Tris-HCl (pH 8.3) and were eluted with this buffer containing 0.10 *M*, 0.25 *M*, 0.50 *M*, and 3.0 *M* NaCl. The percentage of protein eluted in each peak is shown in the upper panel.

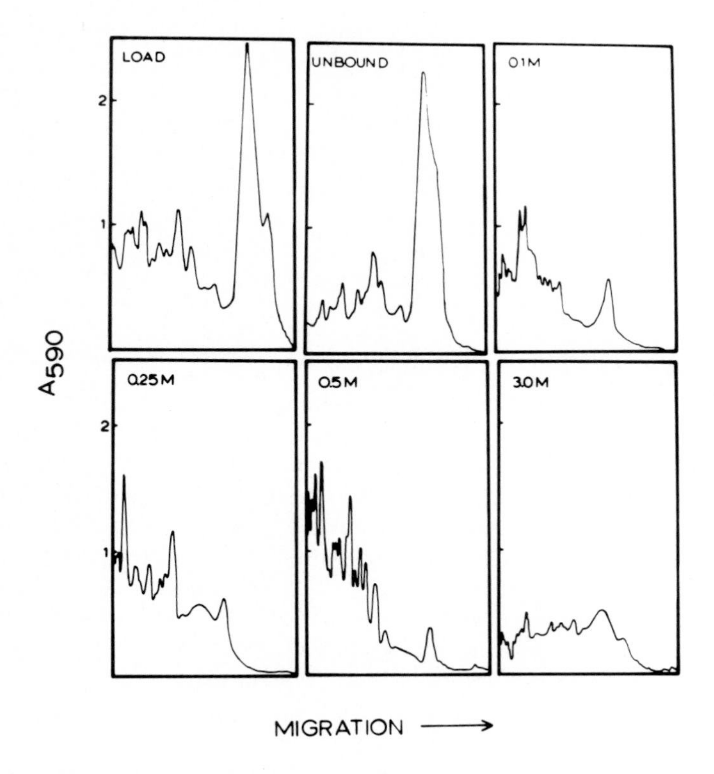

FIGURE 2. SDS-polyacrylamide gel electrophoretic profiles of chromosomal proteins fractionated with QAE-Sephadex®.

ent C_ot values to various types of Sephadex® and Sepharose columns. Nonhistone chromosomal proteins are adsorbed to the DNA in low salt concentrations and a series of different DNA binding protein fractions are eluted with an increasing stepwise salt gradient. Another method which has been used to isolate DNA-nonhistone complexes is nitrocellulose filtration.[113-115]

The hydrophobic nature of several of the nonhistone chromosomal proteins has been utilized as a means of fractionation. A complex group of proteins can be separated from the remaining proteins by affinity chromatography in the presence of high salt concentration to daunomycin-CH-Sepharose 4B.[116] Another method of fractionating a specific group of nonhistone chromosomal proteins is by affinity chromatography to other chromosomal proteins, especially histones.[117] Similar to DNA affinity chromatography, this method has the advantage of separating the nonhistone chromosomal proteins on a functional basis.

One of the most promising approaches for isolation of nonhistone chromosomal proteins which may be involved in genome structure and function is affinity chromatography using cloned DNA. This approach permits preparation of affinity columns with large quantities of DNA sequences representing specific, limited regions of the genome. Gehring and co-workers[118] have successfully utilized affinity columns with cloned Drosophila DNA sequences to isolate specific DNA-binding polypeptides. With increasing availabilty of cloned eukaryotic sequences the broad applicability of affinity chromatography is self evident.

IV. POLYACRYLAMIDE GEL ELECTROPHORETIC TECHNIQUES FOR THE ANALYSIS OF CHROMOSOMAL PROTEINS

Polyacrylamide gel electrophoresis represents the most widely utilized high resolution procedure for quantitative and qualitative analysis of chromosomal polypeptides. Although electrophoretic fractionation procedures are analytical techniques, they can also be used for preparative scale fractionation with subsequent isolation of the fractions. Specific applications of polyacrylamide gel electrophoresis of chromosomal proteins include: (1) to resolve the protein components of chromatin, thus generating a chromosomal protein "fingerprint," and it is this technique that has led to the realization that chromosomal proteins are an extremely complex group comprising at least several hundred different species; (2) to study cell, tissue, or species differences in the complement or postsynthetic modification of chromosomal proteins; (3) to study variations in the complement or postsynthetic modification of these proteins that are associated with development, differentiation, aging, cell cycle traverse, and changes in transcriptional activity and chromatin structure; (4) to identify and quantitate a particular chromosomal protein by its mobility in a gel; (5) to monitor the steps involved in the extraction and purification of chromosomal proteins when, sometimes, no other means are available; and (6) as a means of purification itself. Chromosomal proteins fractionated on polyacrylamide gels can be eluted and subjected to peptide mapping and amino acid composition or sequence analysis. Antibodies can also be raised to proteins eluted from gels.

We will not present a detailed review of the intricacies of polyacrylamide gel electrophoresis as several exist.[119-121] Instead, we will discuss only the most useful systems for the analysis of chromosomal proteins. For detailed descriptions of each method the references quoted in the relevant section should be consulted.

All the slab gel recipes given are for a slab gel apparatus such as that supplied by Bio-Rad® (model 220). However, recipes can easily be adapted for any slab gel apparatus or for cylindrical gels.

A. General Information

1. Washing of Gel Plates and Tubes

Glass plates and tubes are washed in detergent, then soaked in nitric acid followed by extensive rinsing with distilled water. Plates are then rinsed in methanol and dried. Perspex (Plexiglass) plates and tubes are detergent washed, soaked in alcoholic KOH (in KOH diluted 1:1 with methanol) and rinsed and dried as for glass plates.

2. Materials

Acrylamide and N,N′-methylenebisacrylamide are electrophoresis grade. Urea is ultra-pure grade from Schwarz Mann; SDS-specially pure grade from British Drug House (supplied by Gallard-Schlesinger in the U.S.A.); Nonidet 9-40 from Shell Chemical; Ampholytes from LKB, Bio-Rad®, or Pharmacia (because ampholytes from different sources give slightly different results the source of ampholytes should not be varied when comparing samples). Agarose is from Seakem.

B. Analysis of Total Chromosomal Proteins or Nonhistone Chromosomal Proteins

1. One-Dimensional Gel Electrophoretic Systems

Although one-dimensional systems do not provide as much resolution as two-dimensional systems, they have the advantage of being quicker and easier to perform and enable more convenient comparison of multiple samples.

a. SDS-Polyacrylamide Gel Electrophoresis

The first gel electrophoretic analyses of nonhistone chromosomal proteins were made in the absence of detergents but, due to the marked tendency of these proteins to aggregate, very few of the proteins entered the gel matrix. To prevent this problem it is necessary to carry out electrophoresis in the presence of a detergent, usually SDS.

The one-dimensional system that yields the greatest resolution of nonhistone chromosomal proteins is essentially that of Laemmli.[122] This method is based on the discontinuous buffer system of Ornstein[123] and Davis,[124] but with the addition of SDS to the gel and electrophoresis buffers. Due to the denaturing conditions employed, bands on these gels correspond to polypeptide chains and not to native proteins. The distance of migration is inversely proportional to the log of the molecular weight of the polypeptide.[125] However, certain groups of proteins (such as glycoproteins and some histones) deviate slightly from this relationship. SDS-polyacrylamide slab gel electrophoresis is discussed in detail by LeStourgeon and Beyer[126] and Van Blerkom.[127]

To prepare samples use a sample buffer of 0.2% SDS-1% 2-mercaptoethanol-10 m*M* Na phosphate buffer (pH 7.2)-30% glycerol. Before electrophoresis samples should be adjusted to the composition of the sample buffer. If the sample to be analyzed contains a large amount of DNA (for example, chromatin) the sample buffer should contain 10 m*M* EDTA to help the dissociation of proteins from nucleic acids. Samples which have been lyophilized, or precipitated with TCA or alcohol can be resuspended directly in the sample buffer. However, traces of acid should first be removed from the acid-precipitated samples by washing the precipitate twice with absolute ethanol or methanol. Aqueous samples at neutral pH can be directly adjusted to the SDS, 2-mercaptoethanol and glycerol concentrations of the sample buffer by the addition of concentrated stock solutions. However, if the sample has a final salt concentration of more than 200 m*M*, it should be dialysed against the sample buffer as high salt concentrations interfere with electrophoresis. A final protein concentration of ~1 mg/mℓ is ideal. Samples can be stored indefinitely in "sample buffer" at −20° provided they are not repeatedly frozen and thawed.

After solubilization in "sample buffer" place samples in 1.5 mℓ Eppendorf centrifuge tubes and heat at 100° for 3 min. Cool samples to room temperature and add 2-

mercaptoethanol (to replace that lost during heating) and bromophenol blue to final concentrations of 1% and 0.01%, respectively. Centrifuge at 10,000 g for 2 min at room temperature to pellet any particulate materials. Heating the protein samples in SDS and 2-mercaptoethanol ensure their complete dissociation and denaturation of proteins and the reduction of disulfide bridges. This procedure does not cause peptide bond cleavage.[128]

In the preparation of gels, nonhistone chromosomal proteins can be electrophoresed in cylindrical or slab gels, however, slab gels give better resolution and they enable easier and more accurate comparison of multiple samples. Good resolution of nonhistone chromosomal proteins is achieved with a 9 cm 10% polyacrylamide separating gel and a 1 cm 3% polyacrylamide stacking gel. Slabs containing a linear or exponential gradient of polyacrylamide can also be used. See Van Blerkom[127] for methods of making these gradient gels.

The stock solutions are (1) 30% acrylamide: 29.2 g acrylamide, 0.8 g NN′-methylene bisacrylamide. Make up to 100 mℓ with water, filter and store in a dark bottle at 4°. This solution keeps for several months. (2) pH 8.8 buffer: 1.5 *M* Tris-HCl (pH 8.8)-0.4% SDS (18.17 g Tris, 4 mℓ of 10% SDS — adjust to pH 8.8 with 6 *N* HCl and make up to a final volume of 100 mℓ with water). Store at room temperature. (3) pH 6.8 buffer: 0.5 *M* Tris-HCl (pH 6.8)-0.4% SDS (6.06 g Tris, 4 mℓ of 10% SDS — adjust to pH 6.8 and make up to a final volume of 100 mℓ with water). Store at room temperature. (4) TEMED: store at 4°. (5) 1.5% ammonium persulfate which has been made up fresh before use.

In order to prepare the separating gel (9 cm high × 14 cm wide × 0.15 cm thick), mix 10.0 mℓ stock acrylamide, 7.5 mℓ pH 8.8 buffer, and 11 mℓ of water in a side-arm flask, deaerate for 2 min. Add 23 μℓ of TEMED and 1.5 mℓ of ammonium persulfate, mix quickly and pour gel using a propipette attachment. Overlay immediately with 0.1% SDS — this ensures a flat gel surface. Leave for 1 hr to polymerize. The final gel composition is 10% acrylamide-0.375 *M* Tris-HCl (pH 8.8)-0.1% SDS.

To prepare stacking gel first remove overlay solution from separating gel. Mix 1 mℓ of acrylamide stock solution, 2.5 mℓ of pH 6.8 buffer, 6 mℓ water, 10 μℓ TEMED, and 0.5 mℓ ammonium persulfate solution. It is not necessary to deaerate this mixture. Insert well-former and pour solution; leave to polymerize for 1 hr. The final gel composition is 3% acrylamide 0.125 *M* Tris-HCl (pH 6.8)-0.1% SDS.

The gels can be left at room temperature for 1 or 2 days before use provided the surface is kept moist. We find that better resolution of proteins is obtained when the separating gel is left overnight and the stacking gel is poured an hour before use.

The electrophoresis buffer is 0.025 *M* Tris-0.192 *M* glycine-0.1% SDS (6.06 g Tris, 28.8 g glycine, 2 g SDS made up to 2 ℓ with water). Do not titrate this solution with NaOH or HCl. The pH should be about 8.3.

Samples can be loaded with adjustable automatic pipettes. However, with slab gels of 1.5 mm and 0.75 mm thickness the pipette tips cannot reach the bottom of the wells and this results in an increase in sample volume due to mixing with the electrophoresis buffer. This can be prevented by loading the samples with a microsyringe or a drawn out micro-capillary tube. For a complex mixture of proteins such as total nonhistone chromosomal proteins about 30 to 40 μg should be loaded per well. As the complexity of the sample to be analyzed decreases, correspondingly less protein should be loaded.

Electrophoresis samples at 10 mA/slab (for 1.5 mm thick gels) or 20 mA/slab (for 3 mm thick gels) until the tracking dye enters the separating gel. Increase the current to 20 mA or 40 mA/slab, respectively, and continue electrophoresis until the dye has almost reached the bottom of the separating gel.

After electrophoretic separation polypeptides are detected by staining with Coomassie® Brilliant Blue R-250. The stained bands can be recorded by photography or densitometric scanning. For analysis of radiolabeled samples see Section IV.D.

When staining gels, fix and stain protein bands in 0.2% (w/v) Coomassie® Brilliant Blue R-250 in a 45:45:10 (v/v/v) mixture of water, methanol, and acetic acid. Ethanol can be substituted for methanol. Better results are obtained when a high quality dye, such as that supplied by Bio-Rad®, is used. Gels should be stained for at least 2 hr. Destain gels in several changes of a 6:3:1 (v/v/v) mixture of water, methanol, acetic acid. When destaining is almost complete, transfer gels to 7% acetic acid for final destaining and storage. Destaining time is shortened by slowly agitating the gel on a shaking apparatus. Gels can be stored in sealed plastic bags containing a little 7% acetic acid. Place destained gel on an illuminated light box and photograph through an orange filter.

Cut the gel into individual tracks and scan at 590 nm in a spectrophotometer with a gel scanning attachment. The Coomassie® Blue binding capacity of a given protein is more or less linear over a wide range[120] but considerable variation in response can occur between different proteins.[129] As a rough approximation the relative amount of each protein in a sample is proportional to the area under its absorbance peak.

b. Isoelectric Focusing Gel Electrophoresis

Nonhistone chromosomal proteins can also be separated by isoelectric focusing in polyacrylamide gels. This system is not as effective or as commonly used as SDS-polyacrylamide gel electrophoresis. Nevertheless, because isoelectric focusing separates proteins according to their isoelectric points (as opposed to their molecular weight) it can be a useful ancillary technique to the SDS system. The major role of isoelectric focusing is as the first dimension of a two-dimensional gel electrophoretic system.

2. Two-Dimensional Gel Electrophoretic Systems

Nonhistone chromosomal proteins are an extremely complex group of proteins and even the best one-dimensional gel electrophoretic system can resolve only about 100 species. The introduction of two-dimensional systems has greatly increased our ability to resolve these proteins. Although several types of two-dimensional systems have been developed, we will discuss only those that are most useful for the analysis of nonhistone chromosomal proteins.

a. Isoelectric Focusing — SDS-Polyacrylamide Gel Electrophoresis

This system is basically that of O'Farrell[130] except that in the first dimension the acid solution is in the upper electrode chamber. The proteins are separated by isoelectric focusing in the first dimension and SDS-polyacrylamide slab gel electrophoresis in the second dimension. Thus, separation is according to isoelectric point in the first dimension and molecular weight in the second dimension. The method and potential problems are described in more detail by O'Farrell,[130] O'Farrell and O'Farrell,[141] and Van Blerkom.[127]

The major problems involved in analyzing chromosomal proteins by this system are the difficulty in solubilizing all the proteins and the need to remove DNA from the sample before electrophoresis. The first analyses of nonhistone chromosomal proteins by isoelectric focusing relied solely on 8 *M* urea as a denaturant and solubilizing agent. However, this reagent is not sufficient to prevent the aggregation of many of the proteins. The protein aggregates do not enter the gel and can clog the gel pores. The inclusion by O'Farrell[130] of a nonionic detergent in both the gel and the sample buffer went some way towards overcoming this problem. Another advantage of the nonionic detergent is that it permits the inclusion of a small amount (up to 0.25%) of SDS in the sample buffer. The nonionic detergent removes the SDS by forming mixed micelles and thus prevents it from interfering with isoelectric focusing. Nucleic acids can form highly ionic precipitates that bind proteins and hence cause artifacts in separation, although this is alleviated somewhat by loading the sample at the acid end of the gel. The severity of this problem depends on the amount of DNA present, therefore if the sample to be analyzed is, for example, total chromatin, it is necessary to remove the DNA before isoelectric focusing. Several methods can be used, none of which is entirely satisfactory, and it depends on the specific requirements of each experiment as to which is the best. After solubilization of chromatin in 7 *M* urea-3 *M* NaCl the DNA can be removed by centrifugation[130] or by hydroxyapatite.[98] These methods are more suitable for large scale preparations but when it is necessary to analyze multiple samples containing small amounts of material the following enzymic methods are better.

The first method is DNase I digestion[134] modified from Wilson and Spelsberg[43] and O'Farrell.[130] Resuspend chromatin at 0° to a final concentration of ~10 mg DNA/mℓ in DNase buffer: 0.1 *M* Tris-HCl (pH 7.5)-2 m*M* $MgCl_2$-2 m*M* $CaCl_2$. Disperse chromatin by passing repeatedly through a 23 gauge needle. Add an equal volume of DNase buffer containing RNase (5 μg/mg DNA) and DNase I (20 μg/mg DNA). Incubate the suspension at 0° for 1.5 hr while occasionally forcing it through the needle. After incubation, add solid urea to a final concentration of 9 *M* and then dilute 1:1 with 9.5 *M* urea-2% (w/v) Nonidet P-40-2% (w/v) Ampholytes (pH 3.5-10)-5% (v/v) 2-mercaptoethanol. This method removes over 90% of the DNA.

The second method is S_1 nuclease digestion (based on Peterson and McConkey[136]). Resuspend chromatin at room temperature to a final concentration of $20A_{260}$ units/mℓ in 10 *M* urea-0.1% SDS-1 m*M* Tris-HCl (pH 7.4). Add solid urea to bring final urea concentration to 10 *M*. Add 1/10th volume of 0.3 *M* lysine-HCl (pH 3.8)-1% (w/v) SDS-10 *M* urea-25 m*M* $ZnSO_4$. This gives a final pH of 4.2. Add 70 units of S_1 nuclease from *Aspergillus oryzae* (Sigma)[137] per A_{260} unit and incubate at 45° for 5 min. Stop reaction by adding 1/10th volume of 1 *M* Tris-HCl (pH 7.4)-20% Nonidet P-40-1% (w/v) SDS. Add urea, 2-mercaptoethanol, Ampholytes (pH 3.5-10) and Nonidet P-40 to give final concentrations of 10 *M*, 5% (v/v), 2% (w/v) and 2% (w/v), respectively. Note: This system of isoelectric focusing can resolve protein species differing by only a single charge. Therefore, it is important to take precautions against artifactual charge modifications. As far as possible elevated temperatures and long incubations should be avoided and a reducing reagent (such as 2-mercaptoethanol) should be included in all solutions.

To prepare isoelectric focusing gels use the following stock solutions: (1) 30% acrylamide: 28.4 g acrylamide, 1.6 g N,N'-methylene bisacrylamide. Make up to 100 mℓ with water, filter and store at 4°; (2) 10% (w/v) Nonidet P-40: store at 4°; (3) 40% ampholytes: pH 3-10 and pH 5-7, store at 4°; (4) 10% (w/v) ammonium persulfate (freshly prepared); (5) TEMED: store at 4°.

With the following method, the best resolution is obtained with samples that are radiolabeled to a high specific activity with, for example, [^{35}S]-methionine. Small amounts of protein (<20 μg) can be analyzed on a small diameter (2.5 mm) gel. If samples are unlabeled, larger diameter (6 mm) gels must be used to accomodate the increased amount of protein (200 to 400 μg) necessary for detection on the second dimension slab gel. All gels are 12 cm in height.

The recipe given is for 30 mℓ. Seal the ends of cylindrical gel tubes with parafilm or a rubber cap. Weigh 16.58 g. of urea into a side-arm flask, add 4 mℓ of 30% acrylamide solution, 6 mℓ of 10% Nonidet P-40, 5.88 mℓ of water and 1.5 mℓ of pH 3-10 Ampholytes (for a pH 3-10 gradient) or 1.2 mℓ of pH 5-7 Ampholytes plus 0.3 mℓ of pH 3-10 Ampholytes (for a pH 5-7 gradient). Swirl the solution gently until all the urea is dissolved. Deaerate under a vacuum for 1 min. Add 42 μℓ TEMED, 60 μℓ of ammonium persulfate, mix and pipette solution into gel tubes. Overlay the gel mixture with water and leave to polymerize for 1 hr. The final gel composition is 9.2 *M* urea-4% (w/v) Acrylamide-2% (w/v) Nonidet P-40-2% (w/v) Ampholytes.

For loading and isoelectric focusing the samples, the stock solutions are (1) pre-electrophoresis overlay solution: 9.5 *M* urea-2% (w/v) Nonidet P-40-2% (w/v) Ampholytes-5% (v/v) 2-mercaptoethanol. Store in aliquots at −70°; (2) sample overlay solution: 8 *M* urea-1% (w/v) ampholytes. Store in aliquots at −70°; (3) 0.02 *M* NaOH: freshly prepared and degassed before use; (4) 0.01 *M* H_3PO_4. Replace the parafilm on bottom of gels with nylon bolting cloth (held in place by a rubber band). This prevents the gel from slipping during electrophoresis. Remove water from top of gel and overlay gels with 25 μℓ (2.5 mm diameter gels) or 50 μℓ (6 mm diameter gels) of "pre-electrophoresis overlay solution." Fill the gel tubes and upper reservoir with 0.01 *M* H_3PO_4, and the lower reservoir with 0.02 *M* NaOH. Prerun gels for 15 min at 200 v, 30 min at 300 v and 30 min at 400 v (constant voltage), with the cathode on the bottom and the anode at the top. After pre-electrophoresis discard the upper reservoir solution and aspirate liquid from gel surface. Load samples directly on to gel surface. For small diameter (2.5 mm) tubes 25 μℓ is optimal, for large diameter (6 mm) tubes up to 100 μℓ can be loaded. Overlay samples with "sample overlay solution" (10 μℓ for 2.5 mm gels and 25 μℓ for 6 mm gels). Carefully refill the gel tubes and upper reservoir with 0.01 *M* H_3PO_4. Electrophorese at 400 v for 13 hr and 800 v for 1 hr. After isoelectric focusing is complete the gels can either be equilibrated for electrophoresis in the second dimension or analyzed for protein distribution.

In the equilibration of IEF gel for electrophoresis in the second dimension use an equilibration buffer of 2.3% (w/v) SDS-5% (v/v) 2-mercaptoethanol-10% (v/v) glycerol-0.0625 *M* Tris-HCl (pH 6.8). Remove gel from tube. This is most easily done by placing over the end of the gel tube a piece of rubber tubing connected to a syringe and applying gentle pressure to the syringe. Extrude the gel into a screw-capped glass tube containing equilibration buffer. The gel is equilibrated by gentle shaking for 2 hr at room temperature with at least one change of buffer. Gels can be loaded on to the second dimension slab gel or, after only 1 hr of equilibration, they can be stored at −70° for up to several months. Before electrophoresis, frozen gels must be thawed and equilibrated for a further 1 hr in fresh equilibration buffer.

For second dimension SDS-polyacrylamide slab gel electrophoresis, the separation and stacking gels are prepared as described in Section B.1.a except for the following variations. The plates used to construct the slab gel have their top edges bevelled at 45° to cradle the IEF gel. The separating gel is poured to within 1 cm of the top of the plates (giving gel dimensions of 10.5 cm high × 14 cm wide × 0.15 cm thick). A 1 cm high stacking gel is poured about 45 min before loading the IEF gel. For an illustrated explanation of slab gel assembly for 2-D-electrophoresis see O'Farrell,[130] O'Farrell and O'Farrell,[141] and Van Blerkom.[127]

For loading the IEF gel onto the SDS slab gel and for electrophoresis in the second dimension, use stock solutions of 1% agarose in equilibration buffer (store in aliquots at 4°) and electrophoresis buffer (0.025 *M* Tris-0.192 *M* glycine-0.1% SDS, as in Section B.1.a.) Remove overlay from stacking gel. Pipette 1 mℓ of melted agarose solution on to the surface of the stacking gel. Carefully place the IEF gel on to the agarose making sure that there are no air bubbles between the gel and the agarose. For molecular weight estimation a 1 cm section of an IEF gel containing 5 μg/mℓ of each protein molecular weight marker is loaded on to the second dimension slab gel next to the IEF sample gel. When the agarose has set (about 5 min) pour electrophoresis buffer into both chambers and electrophorese as described in Section B.1.a until the second ion front is almost at the bottom of the gel.

To analyze the IEF gels use a staining solution of 0.75 g Coomassie® Brilliant Blue R-250 dissolved in 225 mℓ methanol by stirring for 5 to 10 min. Pour solution into 465 mℓ water and mix thoroughly. Add 22.5 g sulfosalicyclic acid and 75 g TCA with stirring. This solution should be used the same day.

Fix and stain proteins by incubating gels in staining solution at 60° for 30 min. Destain in an 8:3:1 (v/v/v) mixture of water, ethanol, acetic acid. Store gels in 7% acetic acid and scan at 590 nm.

To analyze second-dimension slab gels, stain and destain as described in Section B.1.a. For analysis of radiolabeled proteins see Section D. For 2-D gels it is best to present the results as photographs or autoradiographs. However, the gels or autoradiograms can be scanned by two-dimensional scanners and the information processed by computer techniques (see Allfrey et al.,[138] Garrels,[139] and Bossinger et al.[140]).

b. Nonequilibrium pH Gradient Gel Electrophoresis (NEPHGE)/SDS-Polyacrylamide Gel Electrophoresis (O'Farrell et al.[131])

The NEPHGE/SDS-PAGE system is essentially the same as the IEF/SDS system described above. However, electrophoresis in the first dimension is carried out for a shorter time and consequently proteins do not focus at their isoelectric points (hence nonequilibrium electrophoresis). The advantage of this is that very basic proteins such as histones, with pI's greater than 10, remain on the gel. Thus, this system is particularly useful for analyzing chromosomal proteins since histones and nonhistones can be resolved on the same gel.

NEPHGE/SDS-PAGE is carried out as described for IEF/SDS-PAGE (Section B.2.a) except that NEPHGE gels are not pre-electrophoresed and electrophoresis in the first dimension is at 400 v for only 5 hr.

C. Analysis of Histones

1. One-Dimensional Gel Electrophoretic Systems

a. Acid-Urea Polyacrylamide Gel Electrophoresis

This is the most widely used sysem and involves electrophoresis of histones in 15% polyacrylamide gels in the presence of 0.9 *M* acetic acid-2.5 *M* urea.[142] Separation is dependent on the charge, size, and shape of the molecules and can be achieved on 10 cm gels. Better resolution is obtained on longer gels (up to 30 cm).

Precipitate acid-extracted histones at −20° with two volumes of 95% ethanol. Pellet the precipitate and dry under a stream of nitrogen. Resuspend the pellet in 0.9 *M* acetic acid and add glycerol to a final concentration of 10%.

To prepare gels[142,143] use stock solutions of: (1) 60% acrylamide: 60 g acrylamide, 0.4 g NN′ methylene bisacrylamide made up to 100 mℓ with water. Filter and store at 4°. (At this concentration some of the acrylamide crystallizes at 4°. However, when warmed to room temperature it goes back into solution); (2) 40% acrylamide: 40 g acrylamide, 1.3 g NN′ methylene bisacrylamide made up to 100 mℓ with water as above; (3) 4 *M* urea: 24.02 g made up to a final volume of 100 mℓ with water; (4) 43.2% (v/v) glacial acetic acid in water; (5) 10% ammonium persulfate — freshly prepared; (6) TEMED; (7) electrophoresis buffer: 0.9 *M* acetic acid, pH 2.9 (52 mℓ glacial acetic acid per liter of water).

To form the separating gel (gel dimensions: 9 cm long × 14 cm wide × 0.15 cm thick), mix 6 mℓ of 60% acrylamide solution, 15 mℓ of 4 *M* urea and 3 mℓ of 43.2% acetic acid and deaerate. Add 288 μℓ of 10% ammonium persulfate and 96 μℓ of TEMED. Pour the gel solution, overlay with water and leave for 1 hr to polymerize. The final gel composition will be 15% acrylamide-2.5 *M* urea-0.9 *M* acetic acid.

To prepare the stacking gel (1 cm high) remove overlay solution from separating gel. Mix 2 mℓ of 40% acrylamide solution, 5 mℓ of 4 *M* urea, 1 mℓ of 43.2% acetic acid, 96 μℓ of 10% ammonium persulfate and 32 μℓ TEMED. Insert well-former, pour solution on top of separating gel and allow to polymerize for 1 hr. The final gel composition will be 10% acrylamide-2.5 *M* urea-0.9 *M* acetic acid.

Pre-electrophorese overnight at 50 V in order to remove anionic contaminants that may interfere with the migration of histones. Load samples (up to 2 μg of each histone) as described in Section B.1.a. In a parallel well load methyl green as a tracking dye. The blue component of this dye migrates slightly faster than H4 (the fastest migrating histone). Electrophorese samples towards the cathode at 100 V until the marker-dye has almost reached the bottom of the separating gel (about 7 hr). In this system histones migrate in the order: H4 > H2A > H2B > H3 > H1.

To analyze the gels fix and stain protein bands overnight in 0.1% Amido Black in 40% methanol, 10% acetic acid. Destain gels in 7% acetic acid. Gels can be photographed or scanned at 620 nm as described in Section B.1.a. If protein samples are radiolabeled, gels can be analyzed as described in Section D.

b. Triton-Acid-Urea Gel Electrophoresis

When Triton® X-100 is added to the acid-urea system the order of migration of the histones is altered. The relative mobility of the histones changes as the amounts of Triton® X-100 and urea are varied.[144] Here we describe a system with a final gel composition of 0.4% Triton® X-100-2.5 *M* urea-0.9 *M* acetic acid. The order of histone migration is H1 > H2B > H4 > H3 > H2A. The main advantage of this system is its ability to resolve minor histone species and modified forms of histones. A disadvantage is the possibility of generating artifactual histone heterogeneity due to oxidation of methionine, tryptophan, and cysteine residues; however, this can be prevented by constantly maintaining the histone samples under reducing conditions. For a detailed discussion of this method see Alfageme et al.,[145] Zweidler,[144] and Hardison and Chalkley.[143]

Dissolve histone pellet at a concentration of ~1 mg/mℓ in 2.5 *M* urea-0.9 *M* acetic acid-5% 2-mercaptoethanol. For gel preparation and electrophoresis of samples the stock solutions and methods are essentially the same as those described for acid-urea gels in Section C.1.a except for the following variations: Separating gel (10 cm high × 14 cm wide × 0.15 cm thick). Mix 6 mℓ of 60% acrylamide solution, 18.75 mℓ of 4 *M* urea, 1 mℓ 12% Triton® X-100 and 3.75 mℓ 43.2% acetic acid in a side-arm flask. Deaerate and add 120 μℓ TEMED and 360 μℓ 10% ammonium persulfate. Pour the solution and overlay with 0.4% Triton® X-100, 2.5 *M* urea, 0.9 *M* acetic acid. Leave for 1 hr to polymerize. (1) The final gel composition is 12% acrylamide-2.5 *M* urea-0.9 *M* acetic acid-0.4% Triton® X-100. (2) No stacking gel is used. (3) To prevent the gel from slipping during electrophoresis place nylon bolting cloth over the lower end of gel. (4) A free radical scavenger such as 2-mercapto-ethylamine is pre-electrophoresed through the gel. (5) 0.1% Triton® X-100 is included in the electrode buffer. (6) Sample is loaded directly on to gel surface and then overlaid with electrophoresis buffer. Gels are stained and destained as described for acid-urea gels in Section C.1.a.

c. SDS-Polyacrylamide Gel Electrophoresis

The concentration of acrylamide (10%) used to separate nonhistone chromosomal protein in SDS-polyacrylamide gels is not sufficient to resolve most of the histones. However, if the acrylamide concentration is raised to 15 or 18% histones can be analyzed by this gel system. Although 18% polacrylamide gels afford better resolution of histones, adequate resolution can be obtained with 15% gels (especially if small amounts, 1 or 2 μg of each histone, are loaded). The 15% gels have the advantage of also being able to resolve some of the nonhistone chromosomal proteins, and thus they are very useful in situations where it is necessary to analyze both histones and nonhistone chromosomal proteins on the same gel, such as in the analysis of nucleosomes and nucleosome-associated proteins.[146,147] Prepare samples as described for SDS-polyacrylamide electrophoresis of nonhistone chromosomal proteins (Section B.1.a).

The method for preparing 15% gels is similar to that described for SDS-polyacrylamide gels used to analyze nonhistone chromosomal proteins (Section B.1.a) except for the inclusion of 5 m*M* EDTA in the separating and stacking gel and increased concentrations of Tris and glycine in the electrophoresis buffer.

Use stock solutions of (1) 30% acrylamide: 30 g acrylamide, 0.8 g NN′ methylene bisacrylamide, made to 100 mℓ with water, filtered and stored at 4°. (2) pH 8.8 buffer: 1.5 *M* Tris-HCl (pH 8.8)-0.4% SDS-20 m*M* EDTA (18.17 g Tris, 4 mℓ of 10% SDS solution, 0.74 g EDTA, pH to 8.8 with HCl and make up to a final volume of 100 mℓ with water. (3) pH 6.8 buffer: 0.5 *M* Tris-HCl (pH 6.8)-0.4% SDS-20 m*M* EDTA (6.06 g Tris, 4 mℓ of 10% SDS solution, 0.74 g EDTA, pH to 6.8 with HCl and make up to a final volume of 100 mℓ with water. (4) TEMED. (5) 10% ammonium persulfate, freshly prepared. (6) Electrophoresis buffer: 50 m*M* Tris-0.38 *M* glycine-0.1% SDS. Do not titrate this solution with NaOH or HCl. The pH should be about 8.3.

To prepare the separating gel mix 15 mℓ of stock acrylamide, 7.5 mℓ of pH 8.8 buffer, and 7.25 mℓ of water in a side-arm flask. Deaerate for 2 min and add 25 μℓ TEMED and 250 μℓ ammonium persulfate. Mix and pour. Overlay with 0.1% SDS and leave for 1 hr to polymerize. The final gel composition is 15% acrylamide-0.375 *M* Tris-HCl (pH 8.8)-0.1% SDS-5 m*M* EDTA.

To prepare the stacking gel remove overlay solution from separating gel. Mix 2 mℓ acrylamide stock, 2.5 mℓ pH 6.8 buffer, 5.4 mℓ water, 10 μℓ TEMED and 100 μℓ ammonium persulfate. Insert well-former and pour solution, leave for 1 hr to polymerize. The final gel composition will be 6% acrylamide-0.125 *M* Tris-HCl (pH 6.8)-0.1% SDS-5 m*M* EDTA.

For the preparation of 18% gels, the method is similar to that for 15% gels described above. Acid-extracted histones are not usually contaminated by DNA, therefore EDTA is not required in the gel buffers. Thomas and Kornberg[148,149] report that the following modifications result in improved resolution: the concentration of Tris in the separating gel is increased to 0.75 *M* and the ratio of NN′ methylene bisacrylamide: acrylamide is decreased to 0.15: 30. Although histones can be resolved in a 10 cm gel, better resolution is obtained in 30 cm gels.

The stock solutions to be used are (1) 30% acrylamide: 30 g acrylamide, 0.15 g NN′ methylene bisacrylamide, make up to 100 mℓ with water, filter and store at 4°. (2) pH 8.8 buffer: 3 *M* Tris-HCl (pH 8.8)-0.4% SDS. (3) pH 6.8 buffer: 0.5 *M* Tris-HCl (pH 6.8)-0.4% SDS. (4) TEMED. (5) 10% ammonium persulfate that has been freshly prepared.

For the separating gel, mix 15 mℓ acrylamide, 6.25 mℓ pH 8.8 buffer and 3.5 mℓ water. Deaerate for 2 min, add 10 μℓ TEMED and 250 μℓ ammonium persulfate. Pour solution and overlay with 0.1% SDS. Leave for 1 hr to polymerize. The final gel composition is 18% acrylamide-0.75 *M* Tris-HCl (pH 8.8)-0.1% SDS.

The stacking gel is prepared by first removing overlay solution from separating gel. Mix 1 mℓ acrylamide, 2.5 mℓ pH 6.8 buffer, 6.4 mℓ water, 10 μℓ TEMED, 100 μℓ ammonium persulfate. Insert well-former and pour solution; leave to polymerize for 1 hr. The final gel composition is 3% acrylamide-0.125 *M* Tris-HCl (pH 6.8)-0.1% SDS.

The loading and electrophoresis of samples is carried out as described for SDS-polyacrylamide gel electrophoresis of nonhistone chromosomal proteins (Section B.1.a) except, in order to achieve good resolution of histones on 15% gels it is necessary to electrophorese more slowly using a constant voltage of 35 V until the proteins have entered the separating gel, then 70 V until the tracking dye almost reaches the bottom of the separating gel. The order of histone migration in SDS-polyacrylamide gels is the same as that in acid-urea gels (Section C.1.a): H4 $>$ H2A $>$ H2B $>$ H3 $>$ H1.

In order to analyze gels, they are stained, destained, scanned, and photographed as described for SDS-polyacrylamide gels of nonhistone chromosomal proteins (Section B.1.a), and radiolabeled samples are analyzed as described in Section D.

2. Two-Dimensional Gel Electrophoretic Systems

Histones are a much less complex group of proteins than nonhistone chromosomal proteins, and consequently, it is not usually necessary to resort to two-dimensional gel electrophoretic techniques for their analysis. However, when histones have to be compared very accurately, as for example, in the study of changes in the complement or postsynthetic modification of histones during development/aging, it may be helpful to employ a two-dimensional system. As the three major 2-D gel systems for histone analysis are all based on various combinations of 1-D systems that have already been described, they will be discussed briefly.

In all three systems the histones are separated in the first dimension in a cylindrical gel which is then equilibrated in the appropriate second dimension electrophoresis buffer and then fixed in place on top of the second dimension slab gel (as described in Section B.2.a). The proteins are then electrophoresed through the slab gel (second dimension).

a. Acid-Urea/SDS Polyacrylamide Gel Electrophoresis[150]

Histones are separated in the first dimension in acid-urea gels (see Section C.1.a). The gel is then equilibrated in SDS buffer[130] (Section B.2.a) and the proteins are electrophoresed in the second dimension through an SDS-polyacrylamide slab gel[130,141] (see Section B.2.a). Although the order of histone migration is similar in both dimensions, the separation of various species is increased.

b. Acid-Urea/Triton-Acid-Urea[151,143]

Separation in the first dimension is again in acid-urea gels[142,143] (see Section C.1.a). The gel is then equilibrated in 0.9 *M* acetic acid-1% Triton® X-100. Electrophoresis in the second dimension is through a 1% Triton® X-100-0.9 *M* acetic acid-2.5 *M* urea slab gel (similar to that of Zweidler[144]) (see Section C.1.b). The advantage of this system is that the order of histone migration differs in each dimension and thus the resolution of histones is improved.

c. NEPHGE/SDS Polyacrylamide Gel Electrophoresis[131]

This system has already been described in Section B.2.b. It separates the histones according to different properties (namely, isoelectric point and molecular weight) in each dimension and therefore increases the resolution obtained in 1-D systems. This system can be used to analyze both histones and nonhistone chromosomal proteins.

D. Analysis of Radiolabeled Proteins

The ability to detect and quantitate electrophoretically-separated radiolabeled chromosomal proteins greatly extends the range of research that can be undertaken on the metabolism and modification of these proteins. For example, the rates of synthesis and turnover of various chromosomal proteins can be measured through the use of [^{3}H]- or [^{14}C]-labeled amino acids or [^{35}S]-methionine; and postsynthetic modifications of these proteins such as phosphorylation, acetylation, methylation, and glycosylation (reviewed by Phillips et al.[152]), can be studied through the use of appropriate radiolabeled precursors. If a sample is radiolabeled to a high specific activity, small amounts can be loaded on to the gel and consequently, the protein resolution is increased — this is particularly helpful in two-dimensional systems. After electrophoretic separation of radiolabeled chromosomal proteins in any of the gel systems described, the distribution of radioactivity in the fractionated proteins can be detected by one of the methods described below. If it is required to analyze the distribution of unlabeled protein in addition to that of radiolabel, gels can first be stained and photographed or scanned (as described above) and then subjected to detection of radioactivity.

1. Gel Fractionation and Counting

After protein fixation cut out individual tracks from one-dimensional slab gels and freeze on a block of solid CO_2. Section frozen gels transversely into 1 or 2 mm slices with a razor-blade gel-slicing apparatus (such as that supplied by Bio-Rad®) and place each slice into a scintillation counting vial. Protein spots can also be cut out of 2-D gels. Process gel slices by one of the following methods:

H_2O_2 method — Dry gel slices by incubation at 80° for 2 hr, then dissolve them in 200 $\mu\ell$ of 30% H_2O_2 at 80° for 1.5 hr. This also decolorizes stained slices. Cool to room temperature and add Triton/toluene scintillation fluid (168 mℓ liquifluor [New England Nuclear]/1333 mℓ Triton® X-100/2500 mℓ toluene) (3 mℓ or 10 mℓ for 5 and 20 mℓ vials, respectively).

Tissue solubilizer method — To each slice add 0.3 mℓ of Soluene-350 (Packard) or NCS (Nuclear Chicago) tissue solubilizer and 10 mℓ of scintillation fluid (5 g PPO, 0.5 g POPOP, 1 ℓ toluene). Incubate overnight at room temperature. Measure the radioactivity in each sample (treated by either method) by liquid scintillation spectroscopy.

2. Fluorography

Radiolabeled proteins in gels can be detected directly by fluorography or autoradiography. Scintillation autoradiography (fluorography)[154] is better than autoradiography for detecting ^{3}H-, ^{14}C-, or ^{35}S-labeled proteins. After electrophoresis gels may be stained prior to fluorography, but because Coomassie® blue tends to diffuse during the fluorography washing procedures, amido black must be used (see Section C.1.a). If gels are not stained, the following procedure is used. Wash gel overnight in 40% methanol, 7% acetic acid (with one change of solution), then for 1 hr in 7% acetic acid. Gels can then be treated by either of the following methods:

PPO/DMSO method[154] — Carry out all steps involving DMSO in a fume hood and wear rubber gloves. Incubate gel for 1 hr in DMSO with one change of solution, then for 3 hr (with gentle agitation) in DMSO containing 15% (w/v) PPO. Pour off this solution and wash gel for 1 hr in water (gel can be left overnight in water). Place the gel on a piece of Whatman 3MM filter paper and cover with a sheet of cellulose acetate or nylon bolting cloth. Dry the gel under heat and vacuum in a gel drying apparatus (homemade or commercially supplied). When gel is dry remove acetate or nylon sheet and expose to Kodak® X-Omat R film (or its equivalent) at −70°. The film may be hypersensitized by prefogging.[155] This procedure greatly increases the efficiency of the fluorography and extends the range of linear response of the film.

"Enhance" method — "Enhance" (New England Nuclear) is a commercially available solution for fluorography. Its substitution for PPO/DMSO gives a several-fold enhancement of the fluorographic process. Fix and wash gel as above and incubate in about three volumes of "Enhance" for 1 hr with occasional agitation, then wash gel in water for 1 hr. Dry gel and carry out fluorography as described in the PPO/DMSO method.

3. Autoradiography

Although fluorography is more sensitive for the detection of ^{3}H-, ^{14}C- or ^{35}S- labeled proteins, the process can result in more diffuse bands or spots than are obtained with autoradiography. This can affect resolution (particularly of 2-D gels). Thus, in some cases, where closely spaced proteins on 2-D gels are highly labeled with ^{14}C or ^{35}S, autoradiography may give better results than fluorography. The procedure for autoradiography is the same as that for fluorography except that the gel is dried directly after washing it in acetic acid. Autoradiography is also used to detect ^{32}P-labeled proteins. However, because ^{32}P is a relatively strong Beta-emitter the sensitivity of response can be improved by placing the film between the dried gel and an intensifying screen (as described by Laskey and Mills[156]). A potential problem in the study of ^{32}P-labeled chromosomal proteins (particularly those labeled with ^{32}P orthophosphate) is the possibility of contamination of the proteins with radiolabeled nucleic acid. This can be overcome by hydrolyzing any nucleic acids present in the gel with 5% TCA for 30 min at 90°.[157]

ACKNOWLEDGMENTS

Ian R. Phillips and Elizabeth A. Shephard are supported by the United Kingdom Cancer Research Campaign. Support from the following research grants is acknowledged: PCM77-15947 (National Science Foundation) and 5-217 (National Birth Defects Foundation). We thank Kay Short, June Webster and Vicki Gates for their editorial assistance.

REFERENCES

1. **Behrens, M.**, Untersuchungen an isolierten zell-und Gewebsbestandteilen; isolierung von zellkernen des kalbsherzmuskels, *Physiol. Chem.*, 209, 59, 1932.
2. **Behrens, M.**, *Biochemische Taschenbuch,* Rauen, H., Ed., Springer-Verlag, Berlin, 1956.
3. **Allfrey, V. G., Stern, H., Mirsky, A. E., and Saetren, H.**, Isolation of cell nuclei in non-aqueous media, *J. Gen. Physiol.*, 35, 529, 1952.
4. **Gurney, T. and Foster, D.**, Nonaqueous isolation of nuclei from cultured cells, in *Methods in Cell Biology,* Vol. 16, Academic Press, New York, 1977, 45.
5. **Wray, W., Conn, P. M., and Wray, V. P.**, Isolation of nuclei using hexylene glycol, in *Methods In Cell Biology,* Vol. 16, Academic Press, New York, 1977, 69.
6. **Hershey, A. D. and Dove, W.**, Introduction to lambda, in *The Bacteriophage Lambda,* Cold Spring Harbor Laboratory, Cold Spring Harbor, 1971.
7. **Wilhelm, J. A., Grover, C. M., and Hnilica, L. S.**, Lack of major cytoplasmic protein contamination of rat liver nuclear chromatin, *Experentia,* 28, 514, 1972.
8. **Johns, E. W. and Fomester, S.**, Studies on nuclear proteins. The binding of extra acidic proteins to deoxyribonucleoprotein during the preparation of nuclear proteins, *Eur. J. Biochem.*, 8, 547, 1969.
9. **Bhorjee, J. S. and Pederson, T.**, Nonhistone chromosomal proteins in synchronized Hela cells, *Proc. Natl. Acad. Sci. USA,* 69, 3345, 1972.
10. **Karn, G., Johnson, E. M., Vidali, G., and Allfrey, V. G.**, Differential phosphorylation and turnover of nuclear acidic proteins during the cell cycle of synchronized HeLa cells, *J. Biol. Chem.*, 249, 667, 1974.
11. **Lin, P. P-C., Wilson, R. F., and Bonner, J.**, Isolation and properties of nonhistone chromosomal proteins from pea chromatin, *Mol. Cell Biochem.*, 1, 197, 1973.
12. **Stein, G. and Baserga, R.**, Cytoplasmic synthesis of acidic chromosomal proteins, *Biochem. Biophys. Res. Commun.*, 44, 218, 1971.
13. **O'Malley, B. W., Toft, D. O., and Sherman, M. R. L.**, Progesterone-binding components of chick oviduct. II. Nuclear components, *J. Biol. Chem.*, 246, 1117, 1971.
14. **Carlsson, S.-A., Moore, G. P. M., and Ringertz, N. R.**, Nucleo-cytoplasmic protein migration during the activation of chick erythrocyte nuclei in heterokaryons, *Exp. Cell Res.*, 76, 234, 1973.
15. **Stein, G. S. and Thrall, C. L.**, Evidence for the presence of non-histone chromosomal proteins in the nucleoplasm of HeLa S_3 cells, *FEBS Lett.*, 32, 41, 1973.
16. **Gilmour, R. S. and Paul, J.**, *In Chromosomal Proteins and Their Role in Regulation of Gene Expression,* Stein, G. S. and Kleinsmith, L. J., Eds., Academic Press, New York, 1975, 19.

17. Modak, S. P. and Imaizumi, M.-T., Preparative equilibrium centrifugation of DNA and RNA on Cs_2SO_4-urea gradients, *J. Cell Biol.*, 63, 231a, 1974.
18. Modak, S., Commelin, D., Grosset, L., Imaizumi, M.-T., Monnat, M., and Scherrer, K., DNA synthesis in circulating erythroblasts of anemic duck, *Eur. J. Biochem.*, 60, 407, 1975.
19. Stein, G. S., Stein, J. L., Kleinsmith, L. J., Eds., *Methods in Cell Biology-Chromatin and Chromosomal Protein Research,* Vol. 16, Academic Press, New York, 1977.
20. Stein, G. S., Stein, J. L., and Kleinsmith, L. J., Fractionation of nonhistone chromosomal proteins, in *Methods in Cell Biology,* Stein, G. S., Stein, J. L., and Kleinsmith, L. J., Eds., Academic Press, New York, 1977.
21. Stein, G. S. and Stein, J. L., Phosphorylation of nonhistone chromosomal proteins by nuclear phosphokinases covalently linked to agarose, in *Methods in Cell Biology,* Vol. 19, Stein, G. S., Stein, J. L., and Kleinsmith, L. J., Eds., Academic Press, New York, 1977.
22. Thompson, J. A. and Stein, G. S., Methods for dissociation, fractionation, and selective reconstitution of chromatin, in *Methods in Cell Biology,* Vol. 19, Stein, G. S., Stein, J. L., and Kleinsmith, L. J., Eds., Academic Press, New York, 1977.
23. Johns, E. W., The isolation and purification of histones, in *Methods in Cell Biology,* Vol. 16, Academic Press, New York, 1977, 183.
24. Johns, E. W., Studies on histones: preparative methods for histones fractions from calf thymus, *Biochem. J.*, 92, 55, 1964.
25. Phillips, D. M. P. and Johns, E. W., A fractionation of the histones of group F2a from calf thymus, *Biochem. J.*, 94, 127, 1965.
26. Stellwagen, R. H., Reid, B. R., and Cole, R. D., Degradation of histone during the manipulation of isolated nuclei and deoxyribonucleoprotein, *Biochim. Biophys. Acta,* 155, 581, 1968.
27. Van Der Westhuyzen, D. R. and Von Holt, C., A new procedure for the isolation and fractionation of histones, *FEBS Lett.*, 14, 333, 1971.
28. Bohm, E. L., Strickland, W. N., Strickland, M., Thwaits, B. H., van der Westhuyzen, D. R., and Von Holt, C., Purification of the five main calf thymus histone fractions by gel exclusion chromatography, *FEBS Lett.*, 34, 217, 1973.
29. van der Westhuyzen, D. R., Bohm, E. L., and von Holt, C., Fractionation of chicken erythrocyte whole histone into the six main components by gel exclusion chromatography, *Biochim. Biophys. Acta,* 359, 341, 1974.
30. Strickland, W. N., Brandt, W. F., and von Holt, X., unpublished observation.
31. Brandt, W. F., Bohm, L., and von Holt, C., Proteolytic degradation of histones and site cleavage in histone F2al and F3, *FEBS Lett.*, 51, 88, 1975.
32. Von Holt, C. and Brandt, W. F., Fractionation of histones on molecular sieve matrices, in *Methods in Cell Biology,* Vol. 16, Academic Press, New York, 1977, 205.
33. Frenster, J. H., Allfrey, V. G., and Mirsky, A. E., Repressed and active chromatin isolated from interphase lymphocytes, *Proc. Natl. Acad. Sci. USA,* 50, 1026, 1963.
34. Dingman, W. C. and Sporn, M. B., Studies on chromatin. I. Isolation and characterization of nuclear complexes of deoxyribonucleic acid, ribonucleic acid, and protein from embryonic and adult tissues of the chicken, *J. Biol. Chem.*, 239, 3483, 1964.
35. Frenster, J. H., Nuclear polyanions as derepressors of synthesis of ribonucleic acid, *Nature (London),* 206, 680, 1965.
36. Chonda, S. X. and Cherion, M. G., Isolation and partial characterization of a mercury-binding nonhistone protein component from rat kidney nuclei, *Biochem. Biophys. Res. Commun.*, 50, 1013, 1973.
37. Benjamin, W. B. and Gelhorn, A., Acidic proteins of mammalian nuclei: isolation and characterization, *Proc. Natl. Acad. Sci. USA,* 59, 262, 1968.
38. Jungmann, R. A. and Schweppe, J. S., Binding of chemical carcinogens to nuclear proteins of rat liver, *Cancer Res.*, 32, 952, 1972.
39. Marushige, K., Brutlag, D., and Bonner, J., Properties of chromosomal nonhistone protein of rat liver, *Biochemistry,* 7, 3149, 1968.
40. Elgin, S. C. R. and Bonner, J., Limited heterogeneity of the major nonhistone chromosomal proteins, *Biochemistry,* 9, 4440, 1970.
41. Seale, R. L. and Aronson, A. I., Chromatin-associated proteins of the developing sea urchin embryo. I. Kinetics of synthesis and characterization of non-histone proteins, *J. Mol. Biol.*, 75, 633, 1973.
42. Spelsberg, T. C., Steggles, A. W., Chytil, F., and O'Malley, B. W., Progesterone-binding components of chick oviduct. V. Exchange of progesterone-binding capacity from target to nontarget tissue chromatins, *J. Biol. Chem.*, 247, 1368, 1972.
43. Wilson, E. M. and Spelsberg, T. C., Rapid isolation of total acidic proteins from chromatin of various chick tissues, *Biochim. Biophys. Acta,* 322, 145, 1973.

44. **Yeoman, L. C., Taylor, C. W., Jordan, J. J., and Busch, H.,** Two-dimensional polyacrylamide gel electrophoresis of chromatin proteins of normal rat liver and Novikoff hepatoma ascites cells, *Biochem. Biophys. Res. Commun.*, 53, 1067, 1973.
45. **Steele, W. J. and Busch, H.,** Acidic nuclear proteins of the Walker tumor and liver, *Cancer Res.*, 23, 1153, 1963.
46. **Shelton, K. R. and Allfrey, V. G.,** Selective synthesis of a nuclear acidic protein in liver cells stimulated by cortisol, *Nature (London)*, 228, 132, 1970.
47. **Teng, C. T., Teng, C. S., and Allfrey, V. G.,** Species-specific interactions between nuclear phosphoproteins and DNA, *Biochem. Biophys. Res. Commun.*, 41, 690, 1970.
48. **Teng, C. S., Teng, C. T., and Allfrey, V. G.,** Studies of nuclear acidic proteins. Evidence for their phosphorylation, tissue specificity, selective binding to deoxyribonucleic acid, and stimulatory effects on transcription, *J. Biol. Chem.*, 246, 3597, 1971.
49. **Shelton, K. R. and Neelin, J. M.,** Nuclear residual proteins from goose erythroid cells and liver, *Biochemistry*, 10, 2342, 1971.
50. **Shelton, K. R., Seligy, V. L., and Neelin, J. M.,** Phosphate incorporation into "nuclear" residual proteins of goose erythrocytes, *Arch. Biochem. Biophys.*, 153, 375, 1972.
51. **Shelton, K. R.,** Plasma membrane and nuclear proteins of the goose erythrocyte, *Can. J. Biochem.*, 51, 1442, 1973.
52. **LeStourgeon, W. M. and Wray, W.,** Extraction and characterization of phenolsoluble acidic nuclear proteins, in *Acidic Proteins of the Nucleus*, Cameron, I. L. and Jeter, Jr., J. R., Eds., Academic Press, New York, 1974, 59.
53. **LeStourgeon, W. M. and Rusch, H. P.,** Localization of nucleolar and chromatin residual acidic protein changes during differentiation in *Physarum polycephalum, Arch. Biochem. Biophys.*, 155, 144, 1973.
54. **Spelsberg, T. C., Mitchell, W. M., Chytil, F., Wilson, E. M., and O'Malley, B. W.,** Chromatin of the developing chick oviduct: changes in the acidic proteins, *Biochim. Biophys. Acta*, 312, 765, 1973. 1973.
55. **Stein, G. and Baserga, R.,** The synthesis of acidic nuclear proteins in the prereplicative phase of the isoproterenol-stimulated salivary gland, *J. Biol. Chem.*, 245, 6097, 1970.
56. **Patel, G. and Wang, T. Y.,** Chromatography and electrophoresis of nuclear soluble proteins, *Exp. Cell Res.*, 34, 120, 1964.
57. **Wang, T. Y.,** The isolation, properties, and possible functions of chromatin acidic proteins, *J. Biol. Chem.*, 242, 1220, 1967.
58. **Shaw, L. M. J. and Huang, R. C.,** A description of two procedures which avoid the use of extreme pH conditions for the resolution of components isolated from chromatins prepared from pig cerebellar and pituitary nuclei, *Biochemistry*, 9, 4530, 1970.
59. **Graziano, S. L. and Huang, R. C. C.,** Chromatographic separation of chick brain chromatin proteins using a SP-Sephadex column, *Biochemistry*, 10, 4770, 1971.
60. **van den Broek, H. W. J., Nooden, L. D., Sevall, J. S., and Bonner, J.,** Isolation, purification, and fractionation of nonhistone chromosomal proteins, *Biochemistry*, 12, 229, 1973.
61. **Hill, R. J., Poccia, D. L., and Doty, P.,** Towards a total macromolecular analysis of sea urchin embryo chromatin, *J. Mol. Biol.*, 61, 445, 1971.
62. **Tuan, D., Smith, S., Folkman, J., and Merler, E.,** Isolation of the non-histone proteins of rat Walker carcinoma 256. Their association with tumor angiogenesis, *Biochemistry*, 12, 3159, 1973.
63. **Arnold, E. A. and Young, K. E.,** Isolation and partial electrophoretic characterization of total protein from non-sheared rat liver chromatin, *Biochim. Biophys. Acta*, 257, 482, 1972.
64. **Levy, S., Simpson, R. T., and Sober, H. H.,** Fractionation of chromatin components, *Biochemistry*, 11, 1547, 1972.
65. **Augenlicht, L. H. and Baserga, R.,** Preparation and partial fractionation of nonhistone chromosomal proteins from human diploid fibroblasts, *Arch. Biochem. Biophys.*, 158, 89, 1973.
66. **Bekhor, I., Kung, G. M., and Bonner, J.,** Sequence-specific interaction of DNA and chromosomal protein, *J. Mol. Biol.*, 39, 351, 1969.
67. **Umanskii, S. R., Tokarskaya, V. I., Zotova, R. N., and Migushina, V. L.,** Isolation and heterogeneity of nonhistone proteins of the rat liver chromatin, *Molek. Biol. (Moscow)*, 5, 215, 1971.
68. **Richter, K. H. and Sekeris, C. E.,** Isolation and partial purification of non-histone chromosomal proteins from rat liver, thymus and kidney, *Arch. Biochem. Biophys.*, 148, 44, 1972.
69. **Chaudhuri, S.,** Fractionation of chromatin nonhistone proteins, *Biochim. Biophys. Acta*, 322, 155, 1973.
70. **Monahan, J. J. and Hall, R. H.,** Fractionation of chromatin components, *Can. J. Biochem.*, 51, 709, 1973.
71. **Yoshida, M. and Shimura, K.,** Isolation of nonhistone chromosomal protein from calf thymus, *Biochim. Biophys. Acta*, 263, 690, 1972.

72. MacGillivray, A. J., Cameron, A., Krauze, R. J., Rickwood, D., and Paul, J., The non-histone proteins of chromatin. Their isolation and composition in a number of tissues, *Biochim. Biophys. Acta,* 277, 384, 1972.
73. MacGillivray, A. J., Carroll, D., and Paul, J., The heterogeneity of the non-histone chromatin proteins from mouse tissues, *FEBS Lett.,* 13, 204, 1971.
74. Rickwood, D. and MacGillivray, A. J., Improved techniques for the fractionation of non-histone proteins of chromatin on hydroxyapatite, *Eur. J. Biochem.,* 51, 593, 1975.
75. Shirey, T. and Huang, R. C. C., Use of sodium dodecyl sulphate, alone, to separate chromatin proteins from deoxyribonucleoprotein of *Arabacia punctulata* from chromatin, *Biochemistry,* 8, 4138, 1969.
76. Elgin, S. C. R. and Bonner, J., Partial fractionation and chemical characterization of the major nonhistone chromosomal proteins, *Biochemistry,* 11, 772, 1972.
77. Gronow, M. and Thackrah, T., The nonhistone nuclear proteins of some rat tissues, *Arch. Biochem. Biophys.,* 158, 377, 1973.
78. Gronow, M., Solubilization and partial fractionation of the sulphur-containing nuclear proteins of hepatoma 233 ascites cells, *Eur. J. Cancer,* 5, 497, 1969.
79. Gronow, M. and Griffiths, G., Rapid isolation and separation of the nonhistone proteins of rat liver nuclei, *FEBS Lett.,* 15, 340, 1971.
80. Gonzalez-Mujica, F. and Mathias, A. P., Proteins from different classes of liver nuclei in normal and thioacetamide-treated rats, *Biochem. J.,* 133, 441, 1973.
80a. Langan, T. A., A phosphoprotein preparation from liver nuclei and its effect on the inhibition of RNA synthesis by histones, in *Regulation of Nucleic Acid and Protein Biosynthesis,* Koningsberger, V. V. and Bosch, L., Eds., Elsevier, Amsterdam, 1967, 233.
81. Gershey, E. L. and Kleinsmith, L. J., Phosphoproteins from calf-thymus nuclei: studies on the method of isolation, *Biochim. Biophys. Acta,* 194, 331, 1969.
81a. Kleinsmith, L. J. and Allfrey, V. G., Nuclear phosphoproteins. I. Isolation and characterization of a phosphoprotein fraction from calf thymus nuclei, *Biochim. Biophys. Acta,* 175, 123, 1969.
82. Platz, R. D., Stein, G. S., and Kleinsmith, L. J., Changes in the phosphorylation of non-histone chromatin proteins during the cell cycle of HeLa S_3 cells, *Biochem. Biophys. Res. Commun.,* 51, 735, 1973.
83. Platz, R. D. and Hnilica, L. S., Phosphorylation of nonhistone chromatin proteins during sea urchin development, *Biochem. Biophys. Res. Commun.,* 54, 222, 1973.
84. Fujitani, H. and Holoubek, V., Similarity of the 0.35M NaCl soluble nuclear proteins and the non-histone chromosomal proteins, *Biochem. Biophys. Res. Commun.,* 54, 1300, 1973.
85. Fujitani, H. and Holoubek, V., Fractionation of nuclear proteins by extraction with solutions of different ionic strength, *Int. J. Biochem.,* 6, 547, 1975.
86. Comings, D. E. and Tack, L. O., Non-histone proteins. The effect of nuclear washes and comparison of metaphase and interphase chromatin, *Exp. Cell Res.,* 82, 175, 1973.
87. Kostraba, N. C., Montagna, R. A., and Wang, T. Y., Study of the loosely bound non-histone chromatin proteins. Stimulation of deoxyribonucleic acid-templated ribonucleic acid synthesis by a specific deoxyribonucleic acid-binding phosphoprotein fraction, *J. Biol. Chem.,* 250, 1548, 1975.
88. Prestayko, A. W., Crane, P. M., and Busch, H., Phosphorylation and DNA binding of nuclear rat liver proteins soluble at low ionic strength, *Biochemistry,* 15, 414, 1976.
89. Comings, D. E. and Harris, D. C., Nuclear proteins. II. Similarity of nonhistone proteins in nuclear sap and chromatin, and essential absence of contractile proteins from mouse liver nuclei, *J. Cell Biol.,* 70, 440, 1976.
90. Bekhor, I., Lapeyre, J.-N., and Kim, J., Fractionation of nonhistone chromosomal proteins isolated from rabbit liver and submandibular salivary gland, *Arch. Biochem. Biophys.,* 161, 1, 1974.
91. Murphy, R. F. and Bonner, J., Alkaline extraction of non-histone proteins from rat liver chromatin, *Biochim. Biophys. Acta,* 405, 62, 1975.
92. Russev, G., Anachkova, B., and Tsanev, R., Fractionation of rat liver chromatin nonhistone proteins into two groups with different metabolic rates, *Eur. J. Biochem.,* 58, 253, 1975.
93. Stein, G. S., Stein, J. L., Park, W. D., Detke, S., Lichtler, A. C., Shephard, E. A., Jansing, R. L., and Phillips, I. R., Regulation of gene expression in HeLa S_3 cells, *Cold Spring Harb. Symp. Quant. Biol.,* 42, 1978.
94. Park, W. D., Stein, G. S., and Stein, J. L., Fractionation and partial characterization of S phase HeLa cell nonhistone chromosomal proteins involved with histone gene transcription, in press.
95. Wang, T. Y. and Johns, E. W., Study of the chromatin acidic proteins of rat liver: heterogeneity and complex formation with histones, *Arch. Biochem. Biophys.,* 124, 176, 1968.
96. Patel, G., Patel, V., Wang, T. Y., and Zobel, C. R., Studies of the nuclear residual proteins, *Arch. Biochem. Biophys.,* 128, 654, 1968.
97. Wang, T. Y., Restoration of histone-inhibited DNA-dependent RNA synthesis by acidic chromatin proteins, *Exp. Cell Res.,* 53, 288, 1968.

98. **MacGillivray, A. J. and Rickwood, D.**, The heterogeneity of mouse-chromatin nonhistone proteins as evidenced by two-dimensional polyacrylamide gel electrophoresis and ion-exchange chromatography, *Eur. J. Biochem.*, 41, 181, 1974.
99. **Gilmour, R. S. and Paul, J.**, RNA transcribed from reconstituted nucleoprotein is similar to natural RNA, *J. Mol. Biol.*, 40, 137, 1969.
100. **MacGillivray, A. J. and Rickwood, D.**, Further characterization of the chromatin non-histone proteins by ion-exchange chromatography and two-dimensional gel electrophoresis, *Biochem. Soc. Trans.*, 1, 686, 1973.
101. **Park, W., Jansing, R., Stein, J., and Stein, G.**, Activation of histone gene transcription in quiescent WI-38 cells or mouse liver by a nonhistone chromosomal fraction from HeLa S_3 cells, *Biochemistry*, 16, 3713, 1977.
102. **Goodwin, G. H., Shooter, K. V., and Johns, E. W.**, Interaction of non-histone chromatin protein (high-mobility group protein 2) with DNA, *Eur. J. Biochem.*, 54, 427, 1975.
103. **Stein, G. S., Park, W. D., and Stein, J.**, Fractionation of nonhistone chromosomal proteins, in *Methods in Cell Biology*, Vol. 17, Stein, G., Stein, J., and Kleinsmith, L. J., Eds., 1978, 293.
104. **Chaudhuri, S., Stein, G., and Baserga, R.**, Binding of chromosomal acidic proteins to DNA and chromatin, *Proc. Soc. Exp. Biol. Med.*, 139, 1363, 1972.
105. **Patel, G. L. and Thomas, T. L.**, Some binding parameters of chromatin acidic proteins with high affinity for deoxyribonucleic acid, *Proc. Natl. Acad. Sci. USA*, 70, 2524, 1973.
106. **Kleinsmith, L. J., Heidema, J., and Carroll, A.**, Specific binding of rat liver nuclear proteins to DNA, *Nature (London)*, 226, 1025, 1970.
107. **Kleinsmith, L. J.**, Specific binding of phosphorylated non-histone chromatin proteins to deoxyribonucleic acid, *J. Biol. Chem.*, 248, 5648, 1973.
108. **Wakabayashi, D., Wang, S., Hord, G., and Hnilica, L. S.**, Tissue-specific nonhistone chromatin proteins with affinity for DNA, *FEBS Lett.*, 32, 46, 1973.
109. **Yamamoto, K. R. and Alberts, B. M.**, *In vitro* conversion of estradiol-receptor protein to its nuclear form: dependence on hormone and DNA, *Proc. Natl. Acad. Sci. USA*, 69, 2105, 1972.
110. **Allfrey, V. G.**, DNA-binding proteins and transcriptional control in prokaryotic and eukaryotic systems, in *Acidic Proteins of the Nucleous*, Cameron, I. L. and Jeter, J. R., Jr., Eds., Academic Press, New York, 1974.
111. **Allfrey, V. G., Inoue, A., Karn, J., Johnson, E. M., and Vidali, G.**, Phosphorylation of DNA-binding nuclear acidic proteins and gene activation in the HeLa cell cycle, *Cold Spring Harb. Symp. Quant. Biol.*, 38, 785, 1974.
112. **Allfrey, V. G., Inoue, A., and Johnson, E. M.**, Use of DNA columns to separate and characterize nuclear nonhistone proteins, in *Chromosomal Proteins and their Role in the Regulation of Gene Expression*, Stein, G. and Kleinsmith, L. J., Eds., Academic Press, New York, 1975, 265.
113. **Catino, J. J., Yeoman, L. C., Mandel, M., and Busch, H.**, Characterization of a DNA binding protein from rat liver chromatin which decreases during growth, *Biochemistry*, 17, 983, 1978.
114. **Patel, G. L. and Thomas, T. L.**, Interactions of a subclass of nonhistone chromatin proteins with DNA, in *Chromosomal Proteins and their Role in the Regulation of Gene Expression*, Stein, G. and Kleinsmith, L., Eds., Academic Press, New York, 1975, 249.
115. **Sevall, J. S., Cockburn, A., Savage, M., and Bonner, J.**, DNA-protein interactions of the rat liver non-histone chromosomal protein, *Biochemistry*, 14, 782, 1975.
116. **Kikuchi, H. and Sato, S.**, Fractionation of nonhistone proteins on a column of daunomycin-CH-Sepharose 4B, *Biochim. Biophys. Acta*, 532, 113, 1978.
117. **Conner, B. J. and Comings, D. E.**, Nuclear proteins. V. Studies of histone-binding proteins from mouse liver by affinity chromatography, *Biochim. Biophys. Acta*, 532, 122, 1978.
118. **Weideli, H., Schedl, P., Artavanis-Tsakonas, S., Steward, R., Yuan, R., and Gehring, W. J.**, Purification of a protein from unfertilized eggs of *Drosophila* with specific affinity for a defined DNA sequence and the cloning of this DNA sequence in bacterial plasmids, *Cold Spring Harbor Symp. Quant. Biol.*, 42, 693, 1978.
119. **Mauer, H. R.**, *Disc Electrophoresis and Related Techniques of Polyacrylamide Gel Electrophoresis*, de Gruyter, Berlin, 1971.
120. **Maizel, J. V., Jr.**, *Methods of Virology*, Vol. 5, Maramasch, K. and Kaprowski, H., Eds., Academic Press, New York, 1971, 179.
121. **Drysdale, J. W.**, Isoelectric focusing in polyacrylamide gel, in *Methods of Protein Separation*, Catsimpoolas, N., Ed., Plenum Press, New York, 1975, 93.
122. **Laemmli, U. K.**, Cleavage of structural proteins during the assembly of the head of bacteriophage T4, *Nature (London)*, 227, 680, 1970.
123. **Ornstein, L.**, Disc electrophoresis. I. Background and theory, *Ann. N.Y. Acad Sci.*, 121, 321, 1964.
124. **Davis, B. J.**, Disc electrophoresis. II. Method and application to human serum proteins, *Ann. N.Y. Acad. Sci.*, 121, 404, 1964.

125. Weber, K. and Osborn, M., The reliability of molecular weight determinations by dodecylsulfate polyacrylamide gel electrophoresis, *J. Biol. Chem.*, 244, 4406, 1969.
126. LeStourgeon, W. M. and Beyer, A. L., The rapid isolation, high resolution electrophoretic characterization and purification of nuclear proteins, in *Methods in Cell Biology*, Vol. 16, Academic Press, New York, 1977.
127. Van Blerkom, J., in Methods for the high resolution analysis of protein synthesis. Application studies of early mammalian development, *Methods in Mammalian Reproduction*, Daniel, J. C., Jr., Ed., Academic Press, New York, 1978, 67.
128. Deutsch, D. G., Effect of prolonged 100°C heat treatment in sodium doecyl sulfate upon peptide bond cleavage, *Anal. Biochem.*, 71, 300, 1976.
129. Fazekas De St. Groth, S., Webster, R. G., and Datyner, A., Two new staining procedures for quantitative estimation of proteins on electrophoretic strips, *Biochim. Biophys. Acta*, 71, 377, 1963.
130. O'Farrell, P. H., High Resolution two dimensional electrophoresis of proteins, *J. Biol. Chem.*, 250, 4007, 1975.
131. O'Farrell, P. Z., Goodman, H. M., and O'Farrell, P. H., High resolution two-dimensional electrophoresis of basic as well as acidic proteins, *Cell*, 12, 1133, 1977.
132. Shaw, L. M. J. and Huang, R. C. C., A description of two procedures which avoid the use of extreme pH conditions for the resolution of components isolated from chromatins prepared from pig cerebellar and pituitary nuclei, *Biochemistry*, 9, 4530, 1970.
133. MacGillivray, A. J. and Rickwood, D., The heterogeneity of mouse-chromatin nonhistone proteins as evidenced by two-dimensional polyacrylamide-gel electrophoresis and ion-exchange chromatography, *Eur. J. Biochem.*, 41, 181, 1974.
134. Phillips, I. R., Isolation, Characterization, and Metabolic Properties of Nuclear Proteins from Various Rat Tissues, Ph.D. thesis, University of London, 1976.
135. Wilson, E. M. and Spelsberg, T. C., Rapid isolation of total acidic proteins from chromatin of various chick tissues, *Biochim. Biophys. Acta*, 322, 145, 1973.
136. Peterson, J. L. and McConkey, E. H., Non-histone chromosomal proteins from HeLa cells. A survey by high resolution, two dimensional electrophoresis, *J. Biol. Chem.*, 251, 548, 1976.
137. Vogt, V. M., Purification and further properties of single-strand specific nuclease from Aspergillus oryzae, *Eur. J. Biochem.*, 33, 192, 1978.
138. Allfrey, V. G., Boffa, L. C., and Vidali, G., *Miami Winter Symp., 15*, Academic Press, New York, 1978.
139. Garrels, J.I., Two dimensional gel electrophoresis and computer analysis of proteins synthesized by clonal cell lines, *J. Biol. Chem.*, 254, 7961, 1979.
140. Bossinger, J., Miller, M. J., Vo, K.-P., Geiduschek, E. P., and Xuong, N.-H., Quantitative analysis of two dimensional electrophoretograms, *J. Biol. Chem.*, 254, 7986, 1979.
141. O'Farrell, P. H. and O'Farrell, P. Z., Two dimensional polyacrylamide gel electrophoretic fractionation, in *Methods in Cell Biology*, Vol. 16, Stein, G., Stein, J., and Kleinsmith, L. J., Eds., Academic Press, New York, 1977, 407.
142. Panyim, S. and Chalkley, R., High resolution acrylamide gel electrophoresis of histones, *Arch. Biochem. Biophys.*, 130, 337, 1969.
143. Hardison, R. and Chalkley, R., Polyacrylamide gel electrophoretic fraction of histones, in *Methods in Cell Biology*, Vol. 17, Stein, G., Stein, J., and Kleinsmith, L. J., Eds., Academic Press, New York, 1978, 235.
144. Zweidler, A., Resolution of histones by polyacrylamide gel electrophoresis in presence of non ionic detergents, in *Methods in Cell Biology*, Vol. 17, Stein, G., Stein, J., and Kleinsmith, L., Eds., Academic Press, New York, 1978, 223.
145. Alfagema, C., Zweidler, A., Mahowald, A., and Cohen, L., Histones of Drosophila embryos electrophoretic isolation and structural studies, *J. Biol. Chem.*, 269, 3729, 1974.
146. Rill, R. L., Shaw, B. R., and van Holde, K. E., Isolation and characterization of chromatin subunits, in *Methods in Cell Biology*, Vol. 18, Stein, G., Stein, J., and Kleinsmith, L., Eds., Academic Press, New York, 1978, 69.
147. Phillips, I. R., Shephard, E. A., Tatcher, W. B., Stein, J. L., and Stein, G. S., Evidence for nonhistone chromosomal protein kinase activity associated with nucleosomes isolated from HeLa S_3 cells, *FEBS Lett.*, 106, 56, 1979.
148. Thomas, J. O. and Kornberg, R. D., An octamer of histones in chromatin and free in solution, *Proc. Natl. Acad. Sci. USA*, 72, 2626, 1975.
149. Thomas, J. O. and Kornberg, R. D., The study of histone-histone associations by chemical cross-linking, in *Methods in Cell Biology*, Vol. 18, Stein, G., Stein, J., and Kleinsmith, L. J., Eds., Academic Press, New York, 1978.
150. Woodland, H. R. and Adamson, E. D., The synthesis and storage of histones during the oogenesis of xenopus laevis, *Dev. Biol.*, 57, 118, 1977.

151. **Spiker, S.**, Expression of parental histone genes in the intergeneric hybrid hexaploide, *Nature (London)*, 259, 418, 1976.
152. **Phillips, I. R., Shephard, E. A., Stein, J. L., and Stein, G. S.**, Role of nonhistone chromosomal proteins in selective gene expression, in *Eukaryotic Gene Regulation*, Vol. 2, Kolodny, G., Ed., CRC Press, Boca Raton, Fla., 1981.
153. **Phillips, I. R., Shephard, E. A., Stein, J. L., Kleinsmith, L. J., and Stein, G. S.**, Nuclear protein kinase activities during the cell cycle of HeLa S_3 cells, *Biochim. Biophys. Acta*, 565, 326, 1979.
154. **Bonner, W. M. and Laskey, R. A.**, A film detection method for tritium-labelled proteins and nucleic acids in polyacrylamide gels, *Eur. J. Biochem.*, 46, 83, 1974.
155. **Laskey, R. A. and Mills, A. D.**, Quantitative film detection of ^{3}H and ^{14}C in polyacrylamide gels by fluorography, *Eur. J. Biochem.*, 56, 335, 1975.
156. **Laskey, R. A. and Mills, A. D.**, Enhanced autoradiographic detection of ^{32}P and ^{125}I using intensifying screens and hypersensitive film, *FEBS Lett.*, 82, 314, 1977.
157. **Bhorjee, J. S. and Pederson, T.**, Chromosomal proteins: tightly bound nucleic acid and its bearing on the measurement of nonhistone protein phosphorylation, *Anal. Biochem.*, 71, 393, 1976.

Chapter 6

MOLECULAR MECHANISMS OF ALTERATIONS OF SOME ENZYMES IN AGING

Jean Claude Dreyfus, Axel Kahn, and Fanny Schapira

TABLE OF CONTENTS

I. Introduction 114

II. Glucose-6-Phosphate Dehydrogenase (G6PD) 115
- A. Techniques 115
  - 1. Purification of G6PD 115
  - 2. Properties of G6PD 115
    - a. Heat Lability 115
    - b. Stability to Proteolysis 115
    - c. Immunological Techniques 116
      - 1. Qualitative Tests 116
      - 2. Quantitative Tests 116
    - d. Electrophoretic Techniques 116
      - 1. Cellulose Acetate and Starch Gel Electrophoresis 116
      - 2. Isoelectric Focusing 116
      - 3. Electrophoresis in Polyacrylamide Gradient Gel 117
      - 4. Electrophoresis in SDS 117
    - e. Specific Staining for G6PD 117
  - 3. Conditions for Modifications of G6PD 117
    - a. The Leukemic Factor 117
    - b. Glucose-6-Phosphate 118
    - c. Enzyme Preparations Able to Transform G6PD into its Hyperanodic Forms 118
- B. Results 119
  - 1. Demonstration of the Post-Translational Modifications 119
    - a. Immunological Modifications 119
    - b. In Vivo Modifications of the Isoelectric Point 119
  - 2. Heat Lability of G6PD inSenescent Cells 120
  - 3. Structural Differences Between G6PD Purified from Red and White Blood Cells 121
  - 4. Hyperanodic Forms of G6PD 121
    - a. Leukemic Factor 122
    - b. Glucose-6-Phosphate at Acidic pH 123
    - c. NADP Modifying Proteins 123
      - 1. Action of Red Cell Membrane Preparations 123
      - 2. Identification of the Product of the Reaction 123
      - 3. Specificity of $NADP^+$ 124
      - 4. Comparison with Derivatives of NAD^+ and $NADP^+$ 124

5. Nature of the Product Bound to the Hyperanodic Forms of G6PD. 124

III. Aldolases in Eye Lens 125
A. Techniques 125
1. Purification of Aldolases 125
B. Results 126

IV. Aldolase B in Senescent Rat Liver 127
A. Techniques 127
B. Results 128

V. Tyrosine Amino Transferase (TAT) in Senescent Rat Liver. 129
A. Techniques 129
B. Results 130

References 132

I. INTRODUCTION

In recent years age-related changes have begun to be identified at the molecular level. Many modifications have been found and interpreted according to various theories. Among the alterations of proteins, those due to post-translational changes have been growing in importance. Their mechanisms have been described by several authors in this volume and in previous reviews.[1-5] In this article we shall restrict ourselves to the description of our own work.

Contributions from our laboratory which will be discussed are the following:

- Mechanisms of post-translational modifications of glucose-6-phosphate dehydrogenase
- Post-translational modifications of aldolase in eye lens
- Study of liver aldolase and tyrosine amino transferase in aging

Main techniques which have been utilized include:

- Immunological techniques, that require the preparation of pure enzyme if monospecific antisera are wanted
- Analytical separation techniques, which include:
- Starch gel and polyacrylamide slab gel electrophoresis
- Isoelectric focusing, in liquid column, or in slab or tube polyacrylamide gels
- Molecular weight analysis techniques which include:
- Sucrose density gradients
- Dextran gel filtration
- Electrophoresis on polyacrylamide gel gradients
- Analysis of subunit size by SDS polyacrylamide gel electrophoresis

For each section we shall describe first the techniques, then the results. Classical techniques will not be described in detail and we shall insist mainly on original contributions.

II. GLUCOSE-6-PHOSPHATE DEHYDROGENASE (G6PD)

A. Techniques

1. Purification of G6PD

The first complete purification of human G6PD is due to Yoshida.[6] The method was improved by specific elution from CM Sephadex® column with the substrate, glucose-6-phosphate[7] or an analogue 6-phosphogluconate.[8]

Our technique[9] used a specific elution with the coenzyme, NADP, which allows to obtain a higher yield, and a more stable enzyme. The technique includes the following steps:

- Washing the cells and lysing with 0.029 saponine in 0.005 *M* sodium phosphate buffer pH 6.4
- Batch of DEAE-Sephadex®, pH 6.4 and precipitation by ammonium sulfate (70% saturation)
- DEAE Sephadex® column chromatography
- CM Sephadex® chromatography with selective elution by NADP

This technique yields an enzyme titrating 190 units/mg with a 44,000-fold purification in red cells and 65% yield. It can be used also to purify the enzyme from white blood cells and platelets.[10] More recently a much faster technique has been devised by de Flora et al.[11] based on the affinity properties of 2′,5′ Adenosine diphosphate for G6PD, using this reagent bound to A-Sepharose.

Preparation of G6PD apoenzyme[12] — Glucose-6-phosphate dehydrogenase apoenzyme was prepared by precipitating pure enzyme three times in a 70% saturated $(NH_4)_2SO_4$ solution (pH 8.0 with solid Tris), then dialyzing it for 48 to 72 hr against 500 volumes 50 m*M* Tris-HCl buffer (pH 8)/2 m*M* β-mercaptoethanol/1 m*M* EDTA/ 5% (v/v) glycerol/2 mg/mℓ Norit. The dialysis buffer was changed at least 5 times. The presence of tightly bound $NADP^+$ or NADPH was checked by fluorometry, according to De Flora et al.[13] after hydrolysis of the enzyme sample by papain (ratio papain:glucose-6-phosphate dehydrogenase = 1:50, w/w) for 2 hr at 50°C.

2. Properties of Glucose-6-Phosphate Dehydrogenase

The study of properties of G6PD, either in crude extracts or in purified preparations, has been strictly codified by WHO.[14] In our work with this enzyme, we have not used all the conditions proposed by WHO, and several others have been added. Kinetic studies (Km for G6P and NADP, use of analogues, pH activity curves) were performed according to WHO directions.

a. Heat Lability

Heat lability curves of G6PD have been widely used in studies on aging, particularly by Holliday's group.[15] We have measured[16,17] heat stability at 52° in 50 m*M* Tris HCl buffer (ph 8) containing 0.2 m*M* NADP/2 m*M* β-mercaptoethanol/1 m*M* EDTA/1 mg/mℓ bovine serum albumin.

b. Stability to Proteolysis[12]

Stability to proteolysis of pure enzyme preparations was achieved using trypsin and pronase (ratio proteolytic enzyme:total proteins ≃ 1:100, w/w). The enzymes were incubated in a 50 m*M* phosphate buffer (pH 8.0 for trypsin and pH 7.4 for pronase)/1 m*M* EDTA/1 m*M* β-mercaptoethanol/0.2 m*M* $NADP^+$/1.25 mg/mℓ bovine serum albumin. Proteolysis was stopped after different incubation times by adding 2 m*M* diisopropylfluorophosphate (for trypsin) and cooling at 0°C; then residual enzyme activity was measured.

In vitro incubation of the cell extracts — The experiments of in vitro incubation of the cell extracts were usually performed at 37°C during 24 or 48 hr in a sodium phosphate buffer 0.05 *M* pH 6.4 containing $NADP^+$ 2.10^{-4} *M* EDTA 10^{-3} *M*, dithiothreitol 10^{-3} *M*, diisopropylfluorophosphate 10^{-3} *M*, ε-aminocaproic acid 10^{-2} *M* and albumin 2 mg/mℓ. The influence of each of these components (protectors of the sulfhydril groups and antiproteolytic substances) was determined by omitting one of them from the incubation mixture.

c. Immunological Techniques

1. Qualitative Tests

Antisera were obtained in rabbits and gave one single precipitation line in the Ouchterlony double diffusion plates with crude or purified extracts of leukocytes, platelets, or red blood cells. The precipitation lines show a complete identity pattern. The immunoprecipitates retained some enzymic activity and could be specifically stained for G6PD activity.

2. Quantitative Tests[18]

We used an electroimmunodiffusion technique. On a glass plate a 1% agarose gel was deposited in a 0.05 *M* sodium barbital HCl pH 8.2 containing the antiserum, NADP 2×10^{-5} *M*, EDTA 10^{-3} *M* and ε-aminocaproic acid or diisopropylfluorophosphate 10^{-3} *M* as antiprotease. The gel is 1.7 mm thick; 5 μℓ of cell extracts were used, and standard curves were made with five dilutions of the extract. After 4 hr of electrophoresis the plates were kept overnight, then the peaks of immunoprecipitate were revealed with the specific staining solution for G6PD.

The surface of the immunoprecipitate peaks was proportional to the quantity of antigen applied to the gel, allowing to measure the immunological reactivity with respect to a control. On a graph the enzymatic activity applied to the gel was plotted on the Y axis and the migration area of the immunoprecipitate peak on the X axis. The slope gives the value of the ratio enzymatic activity/immunological reactivity expressed as a percentage of the figures of the standard set.

The same immunodiffusion technique can be used for other enzymes; we obtained good results with phosphoglucose isomerase (which migrates towards the cathode, while G6PD migrates towards the anode)[6], phosphogluconate dehydrogenase, and, in studies not focused on aging, hexosaminidase A and B.[19]

d. Electrophoretic Techniques

1. Cellulose Acetate and Starch Gel Electrophoresis

Cellulose acetate and starch gel electrophoresis are performed by conventional techniques and will not be described here.

2. Isoelectric Focusing[18]

Electrofocusing was performed in an acrylamide ampholine gel according to Drysdale et al.[20] using a 7.5% acrylamide gel and ampholines covering the pH range between 3.5 and 10. The migration lasted 4 hr. A diluted cell extract of total activity about 0.004 UI in a volume of 5 to 10 μℓ was applied to each of the gels; then 5 μℓ of the cyanhemoglobin solution (50 μg/mℓ) previously purified of any glucose-6-phosphate dehydrogenase by means of a chromatography on DEAE Sephadex® pH 6.4 were added to each of the deposits to be used as a colored marker allowing an exact comparison of the different gels between them.

In some cases glucose-6-phosphate dehydrogenase was eluted from the gels according to two methods. In the first one, one of the gels is stained for the enzymatic activity and the nonstained gels, sown with the same enzymatic extract (but 10 to 50 times more concentrated) are cut into fine slices corresponding to the different active bands of the stained gel. In the second case the acrylamide ampholine gels are directly cut into 2 mm slices below the hemoglobin used as a marker.

The separated pieces of gel are crushed in a Tris HCl buffer 0.05 *M* pH 8.0, albumin 2 mg/mℓ $NADP^+$ 2.10^{-4}, dithiothreitol 10^{-3} *M*, EDTA 10^{-3} *M*, diisopropylfluorophosphate 10^{-3} *M* and left overnight at 4°C. The eluate is separated from the acrylamide pieces by means of filtration under pressure in small syringes in which glass wool was piled at the bottom.

In some experiments the same electrofocusing technique was used in the presence of 8 *M* urea for the study of subunits.

3. Electrophoresis in Polyacrylamide Gradient Gel[12]

It was performed in gradients from 3 to 27% acrylamide at pH 9.0 or 8.8. Plates were equilibrated with the electrophoresis buffer: 20 m*M* Tris/glycine (pH 8.0 or 8.8) containing 0.1 m*M* β-mercaptoethanol, 0.02 m*M* $NADP^+$ and 1 m*M* EDTA. Electrophoresis ran for 18 hr at 2°C with 30 mA/plate (10 × 9 × 3 mm). At pH 8.0 dimers predominated at low enzyme concentration (5 μg/mℓ) while tetramers predominated at higher concentration (1 mg/mℓ). At pH 8.8 only dimers were found.

4. Electrophoresis in SDS

Electrophoresis in SDS was performed according to Weber and Osborn.[21] SDS electrophoresis of subunits were performed in polyacrylamide gradient gels as a modification of Laemmli' technique.[22]

e. Specific Staining for G6PD[18]

After electrophoresis the plate was incubated for various periods of time with the following solution: Tris HCl 0.02 *M* pH 8.0, glucose-6-phosphate 6.10^{-4} *M* $NADP^-$ 2.10^{-4} *M* $MgCl_2$ 0.01 *M* tetrazolium salt MTT 0.2 mg/mℓ phenazine methosulfate 0.1 mg/mℓ. The reaction was stopped in 5% acetic acid (v/v) and a picture of the plate was taken.

3. Conditions for Modifications of G6PD

Three types of conditions were used for modifying glucose-6-phosphate dehydrogenase; a peptide "leukemic factor", glucose-6-phosphate, and enzymes which are able to transform the coenzyme, NADP. All three required the initial presence of NADP; when the enzyme was stripped of its coenzyme, the apoenzyme was not modified under any of the above conditions.

a. The Leukemic Factor[23]

Research on the leukemic factor was prompted by the observation that some leukemic granulocytic cells have modified G6PD forms. The leukemic extracts when incubated with normal G6PD, led to modifications of the enzyme identical to those of the enzyme in the leukemic cells themselves.

Preparation of leukemic "G6PD modifying factors" — The leukemic extracts were prepared from the blood of two patients; one with chronic granulocytic leukemia and the other with acute myelocytic leukemia. G6PD from the leukemic granulocytes of these patients showed very abnormal electrofocusing patterns with active bands of lower isoelectric point (Figure 1). The crude granulocyte extracts were prepared in 5 m*M* sodium phosphate buffer, pH 6.4, then either heated for 1 hr at 60° or boiled for 5 min, and ultrafiltered. The ultrafiltrate was concentrated by lyophilization, and then deposited on the top of a Sephadex® G 25 column (80 × 2 cm) previously equilibrated with 5 m*M* phosphate buffer, pH 6.4. The flow rate was 10 mℓ/hr and fractions of 4 mℓ were collected. Each fraction was individually lyophilized and tested against purified platelet G6PD.

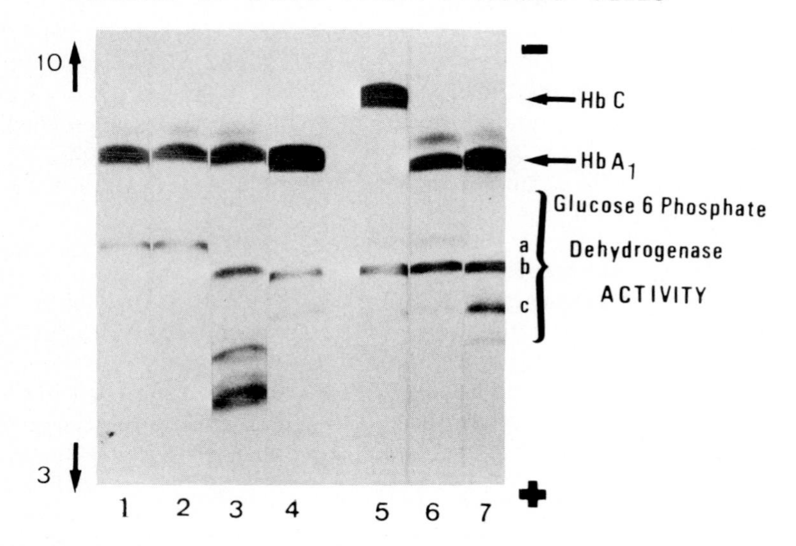

FIGURE 1. Electrofocusing in polyacrylamide gel of glucose-6-phosphate dehydrogenase extracted from various cells. (1) Polymorphonuclear cells; (2) blood platelets; (3) lymphocytes; (4) normal hemolysate; (5) hemolysate from a patient with homozygous hemoglobin C disease; (6) "young" erythrocytes; hemolysate from a control subject; top fraction of red cells separated according to Herz and Kaplan;[23] (7) "old" erythrocytes; hemolysate from a patient with pure red cell anemia. Cell extracts were prepared the day of the experiment. (From Kahn, A., et al., *Biochimie,* 56, 1395, 1974. With permission.)

Studies of the in vitro modifications of platelet G6PD by leukemic "G6PD modifying factors" — Ten microliters of purified G6PD dilution (total activity: 10×10^{-3} IU) were mixed with the factors studied in 50 $\mu\ell$ of 50 m*M* sodium phosphate buffer, pH 6.4, containing 2 mg/mℓ of bovine albumin, 0.2 m*M* $NADP^+$, 1 m*M* dithiothreitol, 1 m*M* EDTA, 10 m*M* ε-aminocaproic acid, and 1 m*M* diisopropylfluorophosphate; then, the preparation was protected from air by a coat of paraffin oil and incubated for 18 or 42 hr at 37°. A 50 $\mu\ell$ sample of the incubated mixture was deposited on the acrylamide-ampholine columns for electrofocusing. In some experiments the influence of pH on the alterations of platelet G6PD by the leukemic "G6PD modifying factors" was studied in Tris chloride buffer pH 8 and pH 9.

b. Glucose-6-Phosphate

Glucose-6-phosphate was incubated in the presence of NADP with pure glucose-6-phosphate dehydrogenase for 6 to 18 hr. Incubation took place usually at pH 6.4 but a whole range of pHs was used. The mechanisms of the action of glucose-6-phosphate were checked by the addition of an NADPH-consuming system, namely a mixture of oxidized glutathione (3 m*M*) and glutathione reductase (10 IU/mℓ).

c. Enzyme Preparations Able to Transform Glucose-6-Phosphate Dehydrogenase into its Hyperanodic Forms

Many tissue preparations can be used. We have been working on two types of material:

Table 1
ENZYMATIC AND IMMUNOLOGICAL ACTIVITY OF GLUCOSE-6-PHOSPHATE DEHYDROGENASE IN THREE HEMOPOIETIC HUMAN CELL

	Specific activity (IU/mg of protein or g of Hb)	Enzymic activity immunological reactivity (% of normal leuko-cytes ratio)	Significance (student's test)
Normal peripheral white blood cells	0.63 ± 0.19	100	
Blood platelets	0.187 ± 0.045	86 ± 6 (13)	$p < 0.01$
Normal erythrocytes — Upper layer	6.3	80 (1)	
Normal erythrocytes — Total	4.8 ± 1	62 ± 6 (13)	$p < 0.001$
Normal erythrocytes — Lower layer	4.0	60 (1)	
Red blood cells			
Blood with a high reticulocyte count (hemolytic anemia with > 300.000 retic/mm³)	8.4 and 8	72 and 89 (2)	
Blood with a low reticulocyte count (marrow-aplasia with < 10.000 retic/mm³	3.8 and 4.7	52 and 53 (2)	

Note: Immunological results are expressed in per cent of those found in freshly prepared white blood cells as standard ± one standard deviation. Numbers in parentheses indicate number of experiments. Upper layer and lower layer of normal erythrocytes were separated by centrifugation in microhematocrit tube. Glucose-6-phosphate dehydrogenase activity is given after substraction of 6-phosphogluconate dehydrogenase activity of that of the two enzymes assayed together.[24]

- A soluble fraction obtained during the purification of leukemic leukocyte extract on CM Sephadex® (in 10 m*M* phosphate pH 6.0) was further purified by DEAE. Sephadex® chromatography.
- Membrane preparations can be obtained from various tissues.

B. Results

1. Demonstration of the Post-Translational Modifications

a. Immunological Modifications[18]

The molecular specific activity of glucose-6-phosphate dehydrogenase was calculated by using the ratio ''enzymatic activity/immunological reactivity'' in leukocytes, platelets, and erythrocytes of 13 different normal bloods. Referred to the leukocytes this molecular specific activity was significantly decreased in platelets (86 ± 6%; $P < 0.01$) and more in the red blood cells (62 ± 6%; $P < 0.001$).

In red blood cells, the ratio ''enzymatic activity/immunological reactivity'' decreased with cell aging as shown (Table 1) by the differences between erythrocytes of two severe hemolytic anemia (in average younger than usual) and erythrocytes of two, not yet transfused, cases of bone marrow failure with a reticulosis which is lower than 10,000/mm³ (cells of an average age greater than normal). A difference of the same kind was found in a normal blood between red cells separated according to their density, the less dense ones being enriched with young cells and reciprocally. In the erythrocytes the decrease of total glucose-6-phosphate dehydrogenase activity with cell aging is due to the decrease of both the number of the active molecules and the molecular specific activity of these molecules.

b. In Vivo Modifications of the Isoelectric Point

Electrofocusing experiments showed that the distribution of the glucose-6-phosphate dehydrogenase active bands varies from a tissue to another and for the same tissue

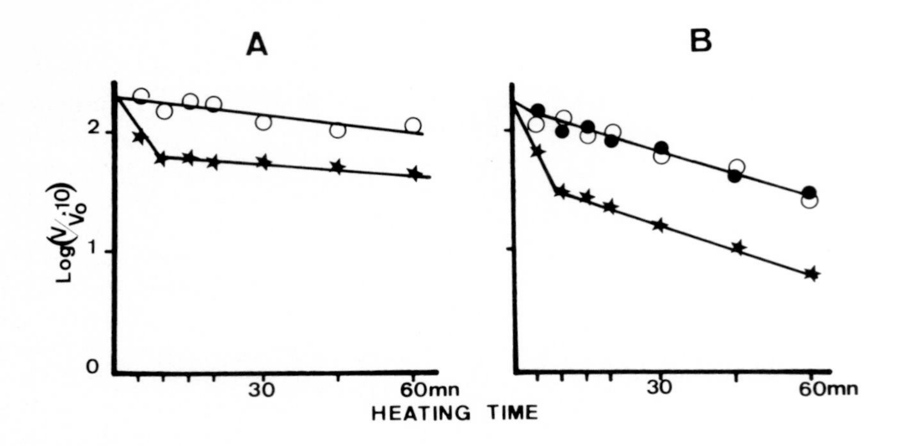

FIGURE 2. Influence of the crude homogenates from old or young fetal lung fibroblasts on the heat lability of exogenous G6PD. Open circles homogenate of cells at the 34th passage; stars, homogenate of cells at the 61st passage; solid circles, pure leukeocyte G6PD. All were diluted in the same Tris-chloride buffer as the cell extracts. (A) Stability of the endogenous enzyme of the cell homogenates. In this experiment serum of a nonimmunized animal was added to the homogenate at the same concentration as that of the antiserum used. (B) Stability of pure leukocyte enzyme added to the homogenates of old or young cells or diluted in the buffer. Endogenous enzyme of the cell homogenates was totally neutralized by the addition of specific antihuman G6PD serum (1 $\mu\ell$/20 units of enzyme activity). After incubation for 1 hr at 37° and 6 hr at 4° the extracts were centrifuged for 30 min at 20,000 g. Then pure human G6PD from leukocytes was added to the cell homogenates in such a manner that the final enzyme activity was identical with that measured before immunoneutralization. (From Kahn, A., et al., *Biochem. Biophys. Res. Commun.*, 77, 760, 1977. With permission.)

with the average age of the enzymatic molecules. Polymorphonuclear cells and platelets showed one active band, clearly predominant, (band a) with a weak band which is more cathodic. The lymphocytes showed a more complex pattern with a main band (band b) which was more anodic than the previous one, with constant position and aspect, and secondary bands of more variable intensity, and clearly more anodic. The hemolysate had two main bands (b and c) and a very weak band "a". Some minor anodic components can also be seen.

In erythrocytes, cell aging goes with a progressive anodization of the bands of glucose-6-phosphate dehydrogenase: preponderance of the band "b" in the hemolysate of blood rich in reticulocytes, disappearance of the band "a" and appearance of a strong band "c", and of anodic components in the hemolysate of an aplasic patient.

2. Heat Lability of Glucose-6-Phosphate Dehydrogenase in Senescent Cells[16-18]

Cells were prepared from liver strain and fetal lung fibroblasts. Most experiments have been performed with liver-derived cells, but results with fibroblasts were similar.

Heat stability of glucose-6-phosphate dehydrogenase purified from old or young cells — We have pooled the cells from different liver-derived strains in phase II and in phase III. The whole homogenate of old cells contained a fraction of heat labile glucose-6-phosphate dehydrogenase amounting to about 35% of the total activity. By contrast, the extract of young cells seemed to contain less than 10% of heat-labile enzyme. After purification, stability of glucose-6-phosphate dehydrogenase from either old or young cells was similar. The yield in enzyme activity of this purification procedure was about 50% for both cell extracts. The same result was obtained with the fetal lung fibroblast strain.

Influence of the crude homogenates from either old or young cells on the stability of exogenous glucose-6-phosphate dehydrogenase — Figure 2 shows the thermal stability curve of pure leukocyte glucose-6-phosphate dehydrogenase added to homogenates of old or young cells whose own enzyme was previously eliminated by specific immunoneutralization: the old cell extracts were able to induce the appearance of a "thermolabile fraction" from exogenous pure glucose-6-phosphate dehydrogenase whereas the young cell extract did not.

3. Structural Differences Between G6PD Purified from Red and White Blood Cells[24]

The enzyme preparations were totally homogeneous as judged by sodium dodecyl-sulfate-acrylamide gel electrophoresis and double immunodiffusion. The molecular weight of the subunits dissociated by sodium dodecyl-sulfate was similar for both preparations. No free N-terminal end could be detected, in any preparation, by the dansylchloride method, in agreement with the data of Yoshida. The treatment of the pure leukocyte enzyme by carboxypeptidase B (Boehringer Mannheim) resulted, upon electrofocusing, in a progressive shift of the G6PD forms towards the acidic pHs. This decrease of isoelectric point was found again as the treated enzyme was dissociated in 8 *M* urea, then focused in gels containing 8 *M* urea. By contrast, carboxypeptidase B remained without any influence on the isoelectric point of erythrocyte G6PD.

The nature of the C-terminal residues of both these G6PD preparations was studied by analyzing the amino acids specifically cleaved by carboxypeptidases A and B. The released residues were either identified by dansylation,[23] or for leukocyte G6PD quantitatively assayed with a Liquimat-F-Labotron amino acid analyzer. The results of these experiments are summarized as follows: Carboxypeptidase A removed a single residue, L-leucine, from leukocyte enzyme, even after a long digestion time. By contrast, glycine and L-alanine were hydrolyzed from red cell G6PD. Since this result is in total agreement with studies previously published by Yoshida,[25] we did not attempt to confirm this finding by quantification of the hydrolyzed residues.

From these results it appeared that pure leukocyte and erythrocyte G6PD differed by their C-terminal ends. The C-terminal residues of the red cell enzyme are Alanine-Glycine as previously reported by Yoshida,[25] whereas the C-terminal end of leukocyte G6PD is Lysine-Leucine.

We have later proved that human platelet G6PD had the same C-terminal end as leukocyte enzyme (i.e., Lysine, Leucine, Bertrand et al., unpublished data).

Structural alterations in vitro of our preparations seem to be rather unlikely since the electrofocusing and electrophoretic patterns, the kinetic properties and the ratio of enzyme activity to G6PD related antigen concentration were not, or only slightly, modified during the purification procedure of either leukocyte or red cell G6PD. Consequently we assume that the structural differences between these enzymes exist in vivo and do not result from in vitro alterations occurring in the course of the purification procedure. Our data suggest that the enzyme found in leukocytes (or in platelets) would be the native G6PD form. The red cell enzyme would arise from this native form by partial proteolysis of Leucine and Lysine. The number, however, of the residues split from the carboxy-terminus must be small since no difference in size can be found between red and white cell enzyme by sodium dodecyl sulfate acrylamide gel electrophoresis. The fact that all the G6PD molecules purified from red cells seem to have a modified C-terminal end could be due to the use, as a starting material for the purification procedure, of old, outdated, blood samples.

4. Hyperanodic Forms of G6PD

A decrease of isoelectric pH was obtained with three kinds of reagents. The resulting modified G6PD was less stable to heat and proteolytic enzymes, and its Km towards G6P was increased.

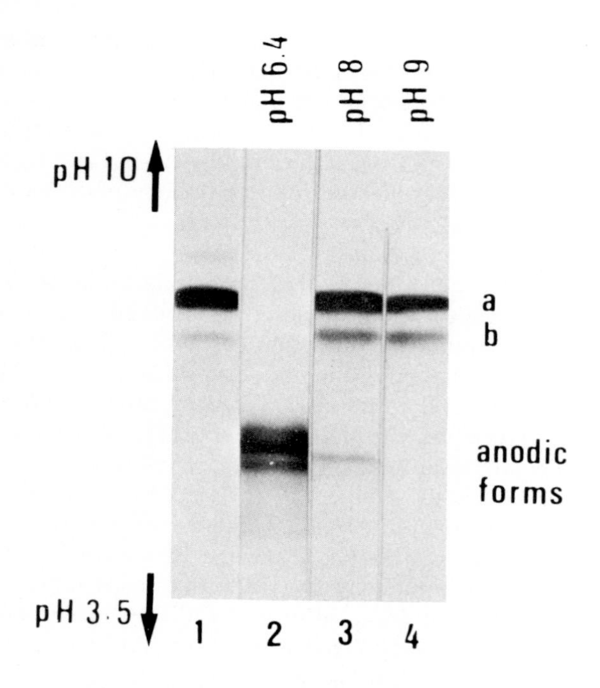

FIGURE 3. Influence of pH on the modifications of purified platelet G6PD by leukemic extracts. The leukemic extracts were boiled and ultrafiltered. (1) Purified platelet G6PD, nonincubated; (2) incubation in the usual conditions (see methods); (3) incubation in 50 m*M* Tris chloride buffer pH 8; (4) incubation in 50 m*M* Tris chloride buffer pH 9; and (5) incubation at pH 6.4 without G6PD modifying factor. (From Kahn, A., et al., *Proc. Natl. Acad. Sci. USA*, 73, 77, 1976. With permission.)

a. Leukemic Factor

Highly purified platelet G6PD showed upon electrofocusing one predominant active band (band a), with a minor more anodic band (band b). When incubated with either the boiled and ultrafiltered leukemic leukocyte extracts or the extracts heated for 1 hr at 60° with 1 m*M* diisopropylfluorophosphate, this pattern was changed into more anodic forms and simultaneously the enzymatic activity decreased 40 to 60%.

By contrast the extracts were left without any influence on platelet G6PD when heated for 1 hr at 60° without diisopropylfluorophosphate. Similarly a previous dialysis abolished the capacity of the extracts to modify platelet G6PD (Table 1).

Figure 3 shows the influence of pH on the modifications of platelet G6PD by the boiled and ultrafiltered extract: the enzyme was modified at pH 6.4, hardly changed at pH 8, and unmodified at pH 9.

After filtration of the boiled, ultrafiltered, and lyophilized extracts on a Sephadex® G 25 column, the fraction able to modify platelet G6PD was found just after the void volume of the column.

The "G6PD modifying factors" were not destroyed by trypsin, chymotrypsin A, elastase, alkaline and acid phosphatases, neuraminidase, carboxypeptidase A, and DNase; by contrast, these factors seemed to be sensitive to pepsin, papain, bromelain, carboxypeptidase B and C, and ribonuclease preparations. Diisopropylfluorophosphate and boiling for 15 min at pH 2.2, however, abolished the effects of the ribonuclease preparations on the G6PD modifying factors, while the RNase activity was not affected by these treatments. Iodoacetamide and iodoacetic acid did not inactivate these factors.

The main characteristics of the modified enzyme were a high Michaelis constant for glucose-6-phosphate (about 100 μM). Molecular specific activity was decreased to 40 to 60% of normal.

b. Glucose-6-Phosphate at Acidic pH

Incubation of pure glucose-6-phosphate dehydrogenase with excess of glucose-6-phosphate (i.e., a ratio of glucose-6-phosphate: $NADP^+ > 2$) caused the total transformation of the native enzyme (pI 6.76) into hyperanodic forms (pI 6.21 to 6.3). At pH 6.4 and 37°C this transformation began after 6 hr of incubation, and was complete after 12 to 18 hr. As with the "leukemic factor", the transition native band a → hyperanodic forms was drastically pH dependent; it took place rapidly at acidic pH (below 6.4), slowly at neutral pH (about 7.0), and not at all above pH 8. There was no means of reversing the glucose-6-phosphate-dependent transformation into hyperanodic forms; especially, the difference of isoelectric point between native enzyme and hyperanodic forms persisted after dissociation and focusing in 8 *M* urea of either untreated preparations or enzyme previously dialysed for 48 hr against 0.5% formic acid.

The glucose-6-phosphate-dependent modification of glucose-6-phosphate dehydrogenase was totally inhibited by an NADPH-consuming system, namely a mixture of oxidized glutathione (3 m*M*) and glutathione reductase (10 I.U./mℓ).

Finally, the incubation of apoenzyme with glucose-6-phosphate, but without $NADP^+$, did not cause transformation into hyperanodic forms.

c. "NADP Modifying Proteins"

Fractions prepared from leukemic leukocyte extracts during G6PD purifications were able to convert G6PD to its hyperanodic forms.[12] Since they have not been yet further characterized, we have later been using cell membranes, and, more specifically, membranes prepared from red blood cells.

1. Action of Red Cell Membrane Preparations[26]

Action of membrane preparations could be shown either with a suspension of membranes or with membranes extracted with 0.5% Triton® X 100. After incubation of G6PD (0.02 units) with 10 or 20 μg of solubilized membranes and 10^{-5} *M* $NADP^+$, native bands have disappeared completely and have been replaced by bands with a lower isoelectric pH.

A direct action of the membrane enzyme on glucose-6-phosphate dehydrogenase was not necessary. When G6PD was placed inside a dialysis bag, and membranes were placed outside in presence of $NADP^+$, the modification of G6PD took place inside the bag. A dialysable product of $NADP^+$ degradation, therefore, was able to react directly with glucose-6-phosphate dehydrogenase.

$NADP^+$ and native G6PD were protected by 10^{-3} *M* nicotinamide as already observed by several authors for reactions involving NAD^+.[27]

By contrast, no reversion could be obtained with nicotinamide, in the presence or absence of the membrane enzyme.

2. Identification of the Product of the Reaction

$NADP^+$ (10^{-4} *M*) was incubated overnight with a membrane preparation. The suspension was then heated to 100°C for 5 min, centrifuged at 10000 g for 10 min, and the supernatant was placed on a Dowex® X2 column (0.5 cm × 18 c) equilibrated with water. Elution was performed in three steps: (1) formic acid 0.2 *M*, (2) formic acid 6 *M*, and (3) formic acid 6 *M* containing 0.8 *M* ammonium formiate. Each peak was concentrated by evaporation and taken up in a small volume of water. The three peaks could be identified: peak 1 as nicotinamide by its migration on PEI and the yellow color of the spot when treated with cyanogen bromide and *p*-aminobenzoic acid; peak 2 as NADP⁺ by migration on PEI plate and specific assay for NADP⁺; peak 3 migrated like an authentic sample of phosphoadenosine diphosphoribose (PADPR) on PEI plate. Each peak was incubated with apo G6PD. Only peak 3 was able to induce anodization of G6PD.

3. Specificity of $NADP^+$

Attempts were made to determine whether $NADP^+$ could be replaced by NAD^+, since it is assumed that the hydrolase can act on NAD^+ and $NADP^+$.[28] This could only be attempted with apo G6PD since traces of $NADP^+$ were enough to bring about the phenomenon.

4. Comparison with Derivatives of NAD^+ and $NADP^+$

Three derivatives of the coenzymes NAD^+ and $NADP^+$ were used: (1) 2′,5′ adenosine diphosphate, which is used in the rapid purification technique of G6PD on account of its affinity towards the enzyme;[11] (2) ADP ribose, a product of the reaction of NAD glucohydrolase on NAD,$^+$ and (3) phospho ADP ribose, a product of the action of the same enzyme on $NADP^+$.

Each of these compounds was incubated with 0.02 UI of pure glucose 6 phosphate dehydrogenase at 10^{-5} to 10^{-3} *M* concentrations for 5 min to 15 hr. Only phospho-ADP ribose was able to modify the isoelectric focusing pattern in the same way as did the membrane enzyme in the presence of $NADP^+$. This effect was observed equally well with the native G6PD as with the stripped apoenzyme. No effect on G6PD was observed with ADP ribose and 2′,5′ ADP.

5. Nature of the Product Bound to the Hyperanodic Forms of Glucose-6-Phosphate Dehydrogenase

An attempt was made to demonstrate binding of a $NADP^+$ derivative to glucose-6-phosphate dehydrogenase. We used the fluorescent analogue of $NADP^+$ *εNADP and a red cell membrane preparation.

εNADP incubated with membranes in the same conditions as that described above for $NADP^+$, gave the same chromatographic profile on the Dowex® X2 column. Peak 3 was fluorescent, and its incubation with G6PD led to the anodization of the enzyme. We conclude that the membranes acted on $\varepsilon NADP^+$ by hydrolyzing the molecule into nicotinamide and εPADP Ribose.

Apo G6PD (200 μg) was incubated with (1) $\varepsilon NADP^+$ 10^{-4} *M* alone, as control; (2) $\varepsilon NADP^+$ 10^{-4} *M* and 50 μℓ of suspended membranes in order to hydrolyse $\varepsilon NADP^+$ into εPADP Ribose and nicotinamide, and (3) $\varepsilon NADP^+$ 10^{-4}*M* and nonfluorescent PADP Ribose 10^{-3} *M*. The pHi of the last two samples was modified by εPADP Ribose and PADP Ribose.

After the incubation, glucose-6-phosphate dehydrogenase was submitted to electrophoresis on cellulose acetate in 6 *M* urea at pH 8.0 for 4 hr in the cold, at a voltage of 135 V. Each strip was cut in two parts; one half was stained for proteins, the other half was eluted in a tris buffer pH 8 and the fluorescence of the eluate was measured in an Aminco Bowman spectrofluorimeter at 300 nm (excitation) and 410 nm (emission). We found that the fluorescence of glucose-6-phosphate dehydrogenase modified by εPADP Ribose was increased four times, compared to fluorescence of the native enzyme. By contrast, no increase in fluorescence was found in the sample incubated with fluorescent $\varepsilon NADP^+$ and nonfluorescent PADP ribose.

The fact that fluorescence persisted after treatment of glucose-6-phosphate dehydrogenase with 6 *M* urea appears as strong evidence for a direct binding of phospho ADP ribose to the enzyme.

* 1 N^6 ethenoadenine dinucleotide phosphate (Sigma).

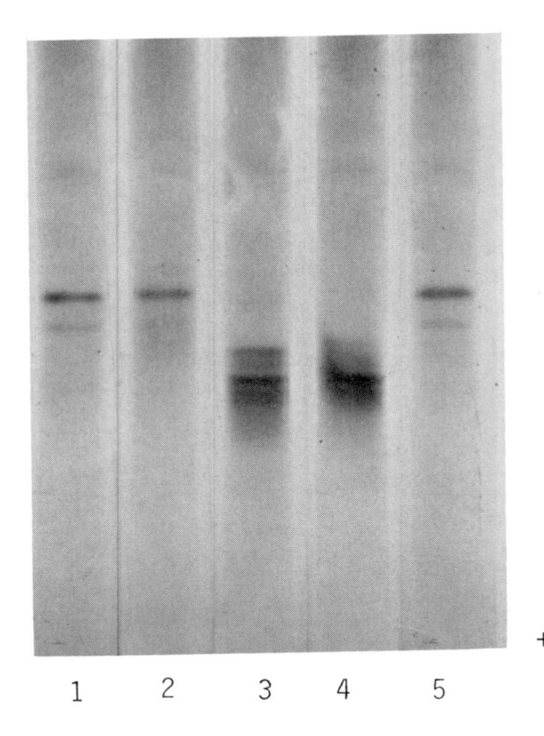

FIGURE 4. Electrofocalization (pH 3-10) on polyacrylamide gel of glucose-6-phosphate dehydrogenase incubated at 37°C overnight with the following compounds. (1) ADP ribose; (2) 2′, 5′ ADP ribose; (3) PADP ribose, (4) peak 3 of the elution on Dowex® of $NADP^+$ treated by membrane enzyme; (5) control. (From Skala, H., et al., *Biochem. Biophys, Res. Commun.*, 89, 988, 1979. With permission.)

III. ALDOLASES IN EYE LENS

Two types of aldolase are found in eye lens; aldolase A (muscle type) and aldolase C (brain type). Aldolase B (liver type) is not present in this organ. We have compared the aldolases of the intermediary zone (cortex fiber cells) and of the central zone (nucleus fiber cells) of the rabbit lens with the same enzyme of the epithelial cells, because protein synthesis is active in the epithelial zone, whereas it is progressively stopping in the intermediary and principally in the central zone. Consequently the eventual enzymic modifications of the central nucleus obligatorily result from post-translational events.

A. Techniques

1. Purification of Aldolases

Aldolase A was prepared from rabbit muscle according to the method of Taylor.[29] Aldolase C was prepared according to the technique of Hatzfeld et al.[30] adapted to rabbit. It includes the following steps:

- Extraction of 20 rabbit brains with 10 m *M* Tris HCL pH 7.5 and EDTA 1 m*M*
- Precipitation by ammonium sulfate between 45 and 65% saturation
- DEAE cellulose chromatography with elution by 0.5 m*M* Fructose 1-6-diphosphate (the specific substrate) and 200 m*M* NaCl

- Electrophoresis on a potato starch block with barbital buffer 50 m*M* pH 8.6
- Isoelectrofocusing on a LKB 8101 column, filled with ampholines 2% (pH 4.0 to 6.0)

Extraction of aldolases — Several eye lenses were removed immediately after killing rabbits and each zone was dissected. Aldolases were extracted with 10 volumes of buffer Tris HCl 10 m*M* pH 7.5

Measure of aldolase activities — FDP activity was measured either according to the spectrophotometric method of Blostein and Rutter[31] or by the colorimetric method of Sibley and Lehninger[32] which we have slightly modified and adapted to use of F1P as substrate (the concentration of F1P was 6.5×10^{-2} *M*[33]).

Electrophoretic technique — Electrophoresis on starch gel was performed in phosphate buffer 0.01 *M* containing β-mercaptoethanol pH 7.0, for 16 hr at 2v/cm. After electrophoresis, the gel was incubated for 1 hr with the following mixture for specific staining of aldolase:

- Tris HCl 0.05 *M* pH 7.5
- Arseniate Na 0.03 *M* pH 7.5
- NAD 0.3 m*M*
- Fructose diphosphate 10 m*M*
- Nitroblue tetrazolium 0.5 mg/mℓ
- Phenazine methosulfate 0.1 mg/mℓ

Immunological techniques — Antisera against both types were prepared in chicken by injecting 1 mg aldolase C in 0.2 mℓ of Freund adjuvant. Injections were repeated each week for 2 months and serum was collected by bleeding the wing vein 10 days after the last injection. Antisera were monospecific, as shown by double diffusion on agar gel. The immunoprecipitates retain some enzymic activity and could be specifically stained by the above technique. In some experiments, extracts were mixed before electrophoresis with an equal volume of antiserum anti-A or anti-C, incubated 1 hr at 4°C and centrifuged at 10,000 g. Supernatants were then submitted to electrophoresis and controls were made with normal chicken serum. Amount of protein antigen was estimated by radial immunodiffusion on plates according to Mancini et al.[34] 100 μℓ of antiserum anti-aldolase C were poured into 4.5 mℓ of agar and 4 μℓ of different dilutions of rabbit lens extracts were put into the wells. Diffusion took place for 4 days and the plates were washed for 2 days. Diameters were measured after specific staining (with the previously described mixture) of the precipitated ring and correction was made for the surface of the well. Surfaces were proportional to the antigen amount, which was expressed in "Arbitrary Units", one unit corresponding to a precipitated area of 100 cm^2 per g of fresh tissue. A ratio F1P aldolase activity (in IU/g)/ Antigen amount (in A.U/g) was determined for each zone; F1P activity was chosen instead of FDP activity because F1P is cleaved to some extent by aldolase C and not (or scarcely) by aldolase A. In these conditions, it was possible to estimate the Aldolase C activity even in hybrids A-C.

B. Results[35]

We shall summarize the results which we obtained. Table 2 shows that in the intermediary and in the central zone, a "Cross-reacting material" for aldolase C does exist; since in these zones no protein biosynthesis occurs, this CRM (that is to say antigenically cross-reactive enzyme molecules, showing that the enzyme becomes partly inactivated) is obligatorily "post-translational".

Table 2
ALDOLASE ACTIVITIES AND AMOUNT OF ANTIGEN IN THE THREE ZONES OF THE LENS

	Aldolase activities (in IU/g)		Antigen amount (in AU/g)	Ratio $\frac{\text{activity}}{\text{Antigen}}$
	FDP	F1P		
Epithelial cells	11.5	0.395	0.116	3.4 ± 0.085
Intermediary zone	32.8	1.07	0.635	1.7 ± 0.078
Central zone	18.3	0.53	0.375	1.4 ± 0.06

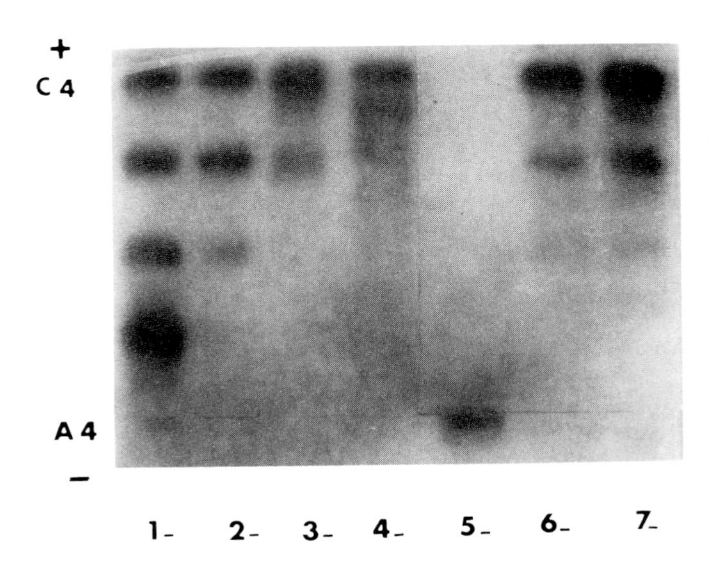

FIGURE 5. (1) Brain, (2) epithelial cells, (3) intermediary zone, (4) central zone, (5) muscle, (6) 6-day-old rabbit lens, (7) 45-day-old rabbit lens. (From Banroques, J., et al., *FEBS Lett.*, 65, 204, 1976. With permission.)

Figure 5 shows three bands in the epithelial zone, the migration of which corresponds to the hybrids A_2C_2 and AC_3, and to the pure tetramer C_4 by comparison with the brain isozymic pattern. The same bands appear in the intermediary and in the central zone, with in addition two supplementary bands: one between A_2C_2 and AC_3, and another, stronger, between AC_3 and C_4. In contrast, no supplementary isozymes appear in the electrophoretic pattern of lens of a 6-day-old rabbit.

Figure 6 shows the electrophoretic pattern after action of antiserum anti-aldolase A and anti-aldolase C: after action of anti-A, the most anodic new hybrid, and the pure tetramer C_4 persists, but the other hybrid disappears. After action of anti-C, all isozymes disappear.

IV. ALDOLASE B IN SENESCENT RAT LIVER

A. Techniques

We used Wistar strain rats, males and females, either 27 to 30 months old, or 3 to 5 months old. Rats were killed by cervical dislocation and exsanguinated; livers were

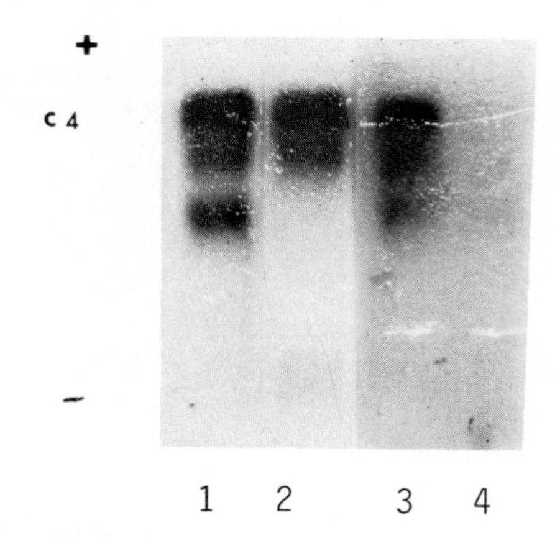

FIGURE 6. Inhibition of rabbit lens aldolase isozymes by antisera (1) and (3) nucleus fiber cells, (2) nucleus fiber cells + anti-aldolase A, (4) nucleus fiber cells + anti-aldolase C. (From Banroques, J., et al., *FEBS Lett.*, 65, 204, 1976. With permission.)

rapidly removed; samples were weighed and then soluble proteins were extracted with Potter Elvejhem apparatus at 4°C with 10 volumes of water and then diluted again with 5 volumes of water.

Purification of rat aldolase B — Extraction was performed in 2 volumes of Tris buffer 0.01 *M* with EDTA 0.001 *M* and β-mercaptoethanol 10 m*M* pH 7.5. Homogenates were centrifuged for 1 hr at 37,000 g. The supernatant was brought to 30% ammonium sulfate saturation, and then centrifuged again for 1 hr at 37,000 g. This supernatant was brought to 45% saturation, and then to 65% using the same procedure. The 65% precipitate was dissolved in the Tris buffer, and submitted to dialysis for 24 hr. The sample was then applied to a Whatman CM 52 column equilibrated with the Tris buffer. Elution was performed with the substrate Fructose di phosphate (2.5 m*M* disolved in Tris buffer pH 7.5). Purification was achieved by isoelectrofocusing (LKB column) in saccharose gradient including 2% ampholines (pH 9.0 to 11.0).

Kinetic methods — Aldolase activities were measured using the colorimetric technique of Sibley and Lehninger adapted by Schapira.[33] (incubation 30 min at 25°C). Apparent Michaelis Constant for the more specific substrate, Fructose-1-phosphate (F1P) was calculated by the Lineweaver Burk method.

Immunological methods — Antiserum to aldolase B was prepared in rabbit, by injecting 0.2 mg in 1 mℓ of phosphate-NaCl buffer with equal volume of complete Freund's adjuvant by intradermal way. Injections were repeated each week for 1 month and serum was collected by bleeding the ear vein 10 days after the last injection. The purity and specificity of antiserum was tested by double immunodiffusion and by immunoelectrophoresis against liver, brain, and muscle extracts. Anti-aldolase B reacts with liver extract, but not with brain and muscle extracts. Antigen amounts were estimated by radial immunodiffusion on plates according to Mancini et al.,[34] with modifications as described above. One "Arbitrary Unit" (A.U.) corresponds to a precipitated surface of 10 cm^2 per g of wet liver.

B. Results[36]

Table 3 shows that the ratio Aldolase activity (in IU): Antigen amount (in AU) does not significantly vary with aging; consequently there was no CRM for aldolase B in senescent rats. Moreover, there are no isozymic modifications, since the aldolase activity ratio FDP:F1P remains constant.

Table 3
ALDOLASE ACTIVITY AND ANTIGEN AMOUNT IN ADULT AND OLD RAT LIVERS

	Adult rats	Old rats
FDP aldolase activity (in IU/g)	8.8 ± 0.5	7.9 ± 0.3
FIP aldolase activity	9.3 ± 0.7	8.2 ± 0.6
FDP/FIP ratio	0.95	0.96
Antigen amount (in AU)	1.26 ± 0.8	1.16 ± 0.04
Aldolase activity / Antigen amount	7.0 ± 0.4	6.5 ± 0.3

The apparent Michaelis Constant of aldolase B for its more specific substrate, F1P, does not vary with aging; 4 m*M* <Km <5 m*M* for both adult and old livers. Similarly thermostability curves were almost identical for adult and old livers.

V. TYROSINE AMINO TRANSFERASE (TAT) IN SENESCENT RAT LIVER

Rats were Wistar strain, 3 to 6 months, and 27 to 31 months old.

A. Techniques

Purification — It was performed according to Belarbi et al.[37]

Preparation of liver extracts — Samples were homogenized with 5 volumes of potassium buffer 50 m*M* pH 6.5 including bovine serum albumin (2 mg/mℓ), dithiothreitol EDTA (1 m*M*). α Ketoglutarate (1 m*M*) and pyridoxal phosphate (0.1 m*M*). Homogenates were centrifuged at 16,000 g. Mitochondrial enzyme is not extracted under these conditions.

TAT activity — It was measured by the Diamonstone method[38] modified by Granner and Tomkins.[39] 1 unit of TAT catalyses the formation of 1 μ*M* of *p*-hydroxyphenyl pyruvate per minute at 37°C.

Immunochemical studies — Antibodies were raised in rabbits which received 1 mg enzyme solution in saline with 0.5 mℓ complete Freund's adjuvant by intradermal injections. Injections (intravenous and intramuscular) were repeated 5 weeks later at 24-hr intervals and blood was taken by heart puncture 12 days after the last injection.[40] Extracts with final identical TAT activities (about 20 mU/mℓ) in extraction buffer were incubated with different amounts of antiserum (diluted in order to obtain 10 to 90% of residual TAT activity in the supernatant after centrifugation). Extracts with antiserum were incubated 1 hr at 37°C and overnight at +4°C and then centrifuged 10′ at 16,000 g. TAT residual activity was measured by incubating the reaction mixture for 30 min at 37°C. Immunoactivation curves are expressed in per cent of residual activity plotted against the antiserum amount.

Study of inducibility "in vivo" — Liver biopsies were performed at 10 a.m. Immediately afterwards, hydrocortisone succinate was injected intraperitoneally (5 mg/100 g body weight). Rats were sacrified 5 hr later.

Isoelectrofocusing — This was performed in 5% polyacrylamide gel containing 2% ampholines pH 3.4 to 10 for 16 hr (1 mA/tube). TAT bands were specifically stained according to the method of Gelehrter et al.[41] with addition of 16 m*M* l-aspartate in order to suppress the aspartate amino transferase band.

Studies of inducibility "in vitro" — Pharmacological quantities of hormones are required for induction in vivo. Consequently we also tested inducibility in vitro using cultured hepatocytes (method of Seglen[42] adapted by Guguen et al.[43]). Liver was per-

Table 4
INDUCIBILITY OF TYROSINE AMINO TRANSFERASE IN VIVO

	TAT activity, IU/g liver	
	Adults rats	Old rats
Before hydrocortisone	3.45 ± 0.4	3.4 ± 0.4
After hydrocortisone	21.8 ± 2.0	32.7 ± 2.4
Mean activity increase	5.7 ± 1.2	10.8 ± 2.6

Note: $p < 0.01$.

Table 5
INDUCIBILITY OF TYROSINE AMINO TRANSFERASE IN VITRO

	TAT activity, IU/g protein	
	Adult hepatocytes	Old hepatocytes
Before dexamethasone	9.0 ± 1.6	7.4 ± 0.9
After dexamethasone	29.9 ± 5.2	21.9 ± 2.4
Mean activity increase	3.6 ± 0.6	3.2 ± 0.4

fused by the portal vein with 200 mℓ of Hepes buffer (acide N-2 hydroxyl ethyl piperazine N′2 ethane sulfonique) pH 7.5 at 37° C (50mℓ/min). Liver was then isolated and perfused by recirculating 100 mℓ of Hepes buffer with 25 mg of Collagenase type I and 5 m*M* CaCL2(solutions saturated with oxygen) for 8 min. The Glisson capsule was removed, and cells were dispersed by agitation. Cell suspension was then filtered on gauze, centrifugated for 45 min at 500 g and washed. Finally cells were suspended in the Eagle medium with addition of 10% fetal calf serum and 1% glucose. Cell viability was estimated by exclusion of trypan blue dye; 3.10^6 viable cells per Falcon flask (25 mℓ) were harvested and incubated at 37°C in 95% air and 4% CO_2 saturated with water. Medium was renewed 5 hr later. Under these conditions the isolated cells were gathering and adhering to the support in 4 hr. Twenty-four hours later, cells formed monolayers; cultures contained only hepatocytes, and not nonparenchymal cells. Dexamethasone (10^{-4} *M*) was added to growth medium for 6 hr after 24 hr of culture.

B. Results[44]

Figure 7 shows the immunoprecipitation curves of liver TAT from adult and old rats using antiserum anti-TAT; it is seen that the same amount of antiserum is required to inactivate the same enzymic activity in both adult and old rats. Consequently, there is no CRM for TAT in "old" TAT. Figure 8 shows the isoelectrofocusing pattern of TAT. Three bands are visible, which seem identical for "young" and "old" TAT. We have concluded that there are no enzymic alterations in TAT of senescent rats; probably because of the very rapid turnover of this enzyme.

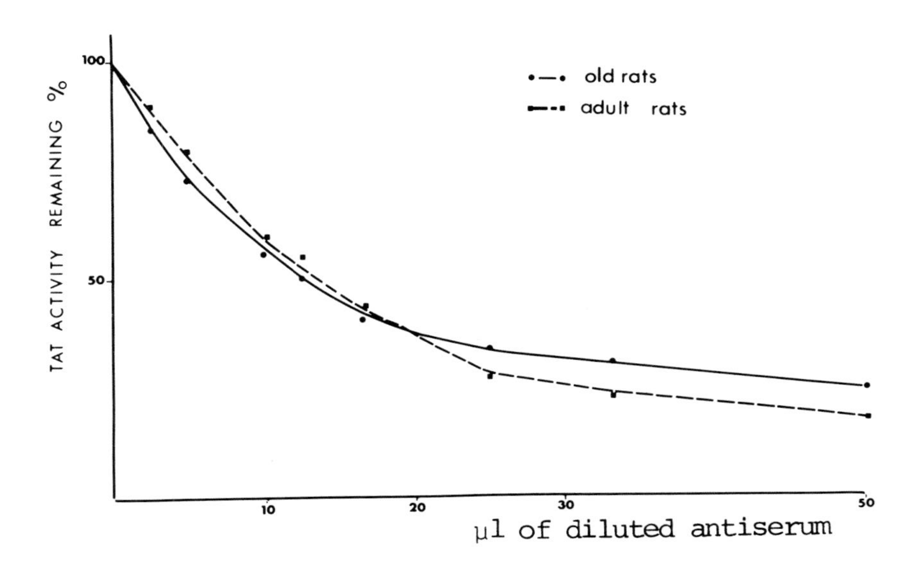

FIGURE 7. Immunoprecipitation of tyrosine amino transferase from livers of adult and old rats. (From Weber, A., et al., *Gerontology,* 126, 9, 1980. With permission.)

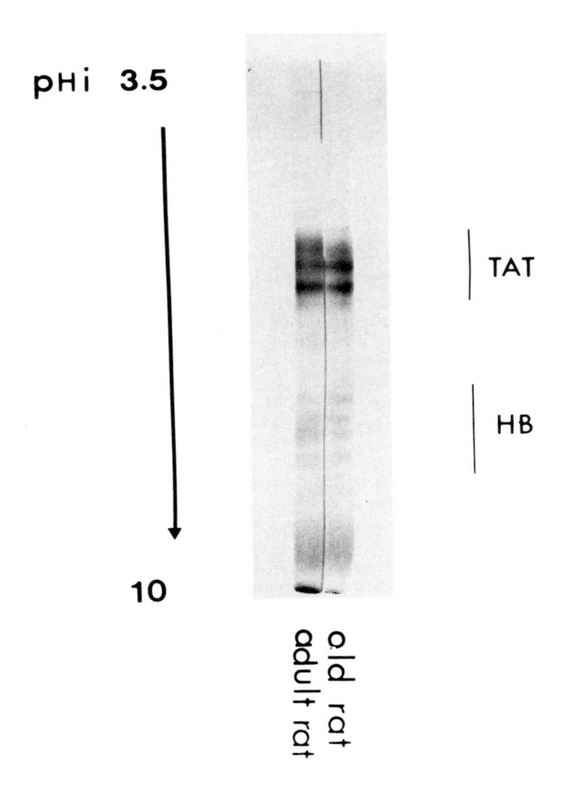

FIGURE 8. Isoelectrofocusing on acrylamide gel of tyrosine amino transferase from livers of adult and old rats. (From Weber, A., et al., *Gerontology,* 126, 9, 1980. With permission.)

REFERENCES

1. Gershon, D., and Gershon, H., An evaluation of the "error catastrophe" theory of aging in light of recent experimental results, *Gerontology,* 22, 212, 1976.
2. Rothstein, M., Aging and alteration of enzymes: a review, *Mech. Ageing Dev.,* 4, 325, 1975.
3. Rothstein, M., Recent developments in age-related alterations of enzymes: a review, *Mech. Ageing Dev.,* 6, 241, 1977.
4. Dreyfus, J. C., Rubinson, H., Schapira, F., Weber, A., Marie, J., and Kahn, A., Possible molecular mechanisms of aging, *Gerontology,* 23, 211, 1977.
5. Dreyfus, J. C., Kahn, A., and Schapira, F., Posttranslational modifications of enzymes, *Curr. Top. Cell Regul.,* 14, 243, 1978.
6. Yoshida, A., Glucose-6-phosphate dehydrogenase of human erythrocytes. Purification and characterization of normal (B) enzyme, *J. Biol. Chem.,* 241, 4966, 1966.
7. Rattazzi, M. C., Isolation and purification of human erythrocyte glucose-6-phosphate dehydrogenase from small amounts of blood, *Biochim. Biophys. Acta,* 181, 1, 1969.
8. Yoshida, A., Enzyme purification by selective elution with substrate analog from ion-exchange columns, *Anal. Biochem.,* 37, 357, 1970.
9. Kahn, A. and Dreyfus, J. C., Purification of glucose-6-phosphate dehydrogenase from red blood cells and from human leukocytes, *Biochim. Biophys. Acta,* 334, 257, 1974.
10. Cottreau, D., Kahn, A., and Boivin, P., Human platelet glucose-6-phosphate dehydrogenase, *Enzyme,* 21, 140, 1976.
11. De Flora, A., Morelli, A., Benatti, V., and Gioliano, F., An improved procedure for rapid isolation of glucose-6-phosphate dehydrogenase from human erythrocytes, *Arch. Biochem. Biophys.,* 169, 362, 1975.
12. Kahn, A., Vibert, M., Cottreau, D., Skala, H., and Dreyfus, J. C., Hyperanodic forms of human glucose-6-phosphate dehydrogenase, *Biochim. Biophys. Acta,* 526, 318, 1978.
13. De Flora, A., Morelli, A., and Giuliano, F., Human erythrocyte glucose-6-phosphate dehydrogenase. Content of bound coenzyme, *Biochem. Biophys. Res. Commun.,* 59, 406, 1974.
14. WHO, Normalisation des Techniques d'Étude de la Glucose 6-Phosphate Deshydrogenase, Genève, 1967.
15. Holliday, R., and Tarrant, G. M., Altered enzymes in aging human fibroblasts, *Nature (London),* 238, 26, 1972.
16. Kahn, A., Guillouzo, A., Cottreau, D., Marie, J., Bourel, M., Boivin, P., and Dreyfus, J. C., Accuracy of protein synthesis and in vitro aging: search for altered enzymes in senescent cultured cells from human livers, *Gerontology,* 23, 174, 1977.
17. Kahn, A., Guillouzo, A., Leibovitch, M. P., Cottreau, D., Bourel, M., and Dreyfus, J. C., Heat lability of glucose-6-phosphate dehydrogenase in some senescent human cultured cells. Evidence for its postsynthetic nature, *Biochem. Biophys. Res. Comm.,* 77, 760, 1977.
18. Kahn, A., Boivin, P., Vibert, M., Cottreau, D., and Dreyfus, J. C., Postranslational modifications of human glucose-6-phosphate dehydrogenase, *Biochimie,* 56, 1395, 1974.
19. Dreyfus, J. C., Poenaru, L., Vibert, M., Ravise, N., and Boue, J., Characterization of a variant of β hexosaminidase "Hexosaminidase Paris", *Am. J. Hum. Genet.,* 29, 287, 1977.
20. Drysdale, J. W., Righetti, P. G., and Bunn, H. F., The separation of human and animal hemoglobins by isoelectric focusing in polyacrylamide gel, *Biochim. Biophys. Acta,* 229, 42, 1971.
21. Weber, K. and Osborne, M., The reliability of molecular weight determination by dodecyl sulfate-polyacrylamide gel electrophoresis, *J. Biol. Chem.,* 244, 4406, 1969.
22. Laemmli, U. K., Cleavage of structural proteins during the assembly of the head of bacteriophage T_4, *Nature (London),* 227, 680, 1970.
23. Kahn, A., Boivin, P., Rubinson, H., Cottreau, D., Marie, J., and Dreyfus, J. C., Modifications of glucose-6-phosphate dehydrogenase and other enzymes by a factor of low molecular weight abundant in some leukemic cells, *Proc. Natl. Acad. Sci. USA,* 73, 77, 1976.
24. Kahn, A., Bertrand, O., Cottreau, D., Boivin, P., and Dreyfus, J. C., Evidence for structural differences between human glucose-6-phosphate dehydrogénase purified from leucocytes and erythrocytes, *Biochem. Biophys. Res. Comm.,* 77, 65, 1977.
25. Yoshida, A., Subunit structure of human glucose-6-phosphate dehydrogenase and its genetic implication, *Biochem. Genet.,* 2, 237, 1968.
26. Skala, H., Vibert, M., Kahn, A., and Dreyfus, J. C., Phospho A.D.P. ribosylation of human glucose-6-phosphate dehydrogenase: probable mechanism of the occurrence of hyperanodic forms, *Biochem. Biophys. Res. Commun.,* 89, 988, 1979.
27. Pekala, P. and Anderson, B. M., Studies of bovine erythrocyte. NAD glycohydrolase, *J. Biol. Chem.,* 253, 7453, 1978.

28. **Honjo, T., Nishizuka, Y., Kato, I., and Hayaishi, O.**, Adenosine diphosphate ribosylation of aminoacyl transferase II and inhibition of protein synthesis by diphteria toxin, *J. Biol. Chem.*, 246, 4251, 1971.
29. **Taylor, J. F.**, Aldolase from muscle, *Methods Enzymol.*, 1, 310, 1975.
30. **Hatzfeld, A., Elion, J., Mennecier, F., and Schapira, F.**, Purification of aldolase C from rat brain and hepatoma, *Eur. J. Biochem.*, 77, 37, 1977.
31. **Blostein, R. and Rutter, W. J.**, Comparative studies of liver and muscle aldolase; immunochemical and chromatographic differentiation, *J. Biol. Chem.*, 238, 1963.
32. **Sibley, J. A. and Lehninger, A. D.**, Determination of aldolase in animal tissues, *J. Biol. Chem.*, 177, 859, 1949.
33. **Schapira, F.**, Dosage des aldolases sériques, *Rev. Et. Clin. Biol.*, 5, 500, 1960.
34. **Mancini, G., Vaerman, J. P., Carbonara, A. O., and Heremans, J. F,.** A single radial immuno diffusion method for the immunological quantitation of proteins, *Protides Biol. Fluids,* 11, 370, 1964.
35. **Banroques, J., Gregori, C., and Schapira, F.**, Post-synthetic modifications of aldolase isozymes in rabbit lens during aging, *FEBS Lett.*, 65, 204, 1976.
36. **Weber, A., Gregori, C., and Schapira, F.**, Aldolase B in the liver of senescent rats, *Biochim. Biophys. Acta,* 444, 810, 1976.
37. **Belarbi, A., Bollack, C., Befort, N., Beck, J. P., and Beck, G.**, Purification and characterization of rat liver tyrosine amino transferase, *FEBS Lett.*, 75, 221, 1977.
38. **Diamondstone, T.**, Assay of tyrosine transaminase activity by conversion of p-hydroxy phenyl pyruvate kinase 10 p-hydroxy benzaldehyde, *Anal. Biochem.*, 16, 395, 1966.
39. **Granner, D. K. and Tomkins, G. M.**, Tyrosine amino transferase (rat liver), *Methods Enzymol.*, 17, 633, 1970.
40. **Beck, J. P., Beck, G., Wong, K. J., and Tomkins, G. M.**, Synthesis of inducible tyronic amino transferase in a cell-free extract from cultured hepatoma cells, *Proc. Natl. Acad. Sci. USA*, 69, 3615, 1972.
41. **Gelehrter, B. O., Emanuel, J. R., and Spencer, C. J.**, Induction of tyrosine transaminase by desamethasone, insuline and serum, *J. Biol. Chem.*, 247, 6197, 1972.
42. **Seglen, P. O.**, Preparation of rat liver cells. II. Effects of ions and chelators on tissue dispersion, *Exp. Cell. Res.*, 76, 25, 1973.
43. **Guguen, C., Guillouzo, A., Boisnard, M., La Cam, A., and Bourel, M.**, Etude ultra structurale de monocouches d'hépatocytes de rat adulte cultivés en présence d'hémisuccinate d'hydrocortisone, *Biol. Gastroenterol.*, 8, 223, 1975.
44. **Weber, A., Guguen-Guillouzo, C., Szajnert, M. F., Beck, G., and Schapira, F.**, Tyrosine amino transferase in senescent rat liver, *Gerontology,* 126, 9, 1980.

Chapter 7

ALTERED POLYPEPTIDE HORMONES AND AGING

Vernon J. Choy

TABLE OF CONTENTS

I. Introduction 136

II. Polypeptide Hormone Polymorphism 136
- A. Categories of Polymorphism 136
- B. Glycoprotein Hormone Polymorphism 139

III. Altered Polypeptide Hormones and Aging 140

IV. Analysis for TSH Polymorphism in Aged Rats 141
- A. Animals 141
- B. Assays 142
  - 1. Preliminary Considerations 142
  - 2. TSH Radioimmunoassay 144
- C. Gel Filtration Chromatography 145
- D. Affinity Chromatography of Glycoproteins 146
  - 1. Introduction 146
  - 2. Procedures 147
- E. Scheme for Analysis of TSH Polymorphism 147
- F. TSH Polymorphism in Aged Rats 149

V. Final Remarks 154

Acknowledgments 155

References 155

I. INTRODUCTION

Since aging can be described in terms of a decline in homeostatic competence of an organism, perhaps due to deterioration in, among other things, synthetic processes, numerous investigations in gerontology have sought to ascertain whether or not abnormal substances accumulate in aged tissue. Provided that misdirected or inappropriate biosynthesis is high in the hierarchy of primary aging mechanisms, it should be possible to detect such dysfunction by the presence of altered forms of quite universal classes of substances such as proteins (see other chapters in this volume). However, error theory,[1,2] which proposes that mistakes arising in the transcription and translation of DNA become self-perpetuating, leading to a gradual but irreversible and accelerating breakdown in the fidelity of protein synthesis, and ultimately leading to error catastrophe, has not been substantiated experimentally. Foote and Stulberg found no age-related change in either the efficiency or fidelity of protein synthesis when tRNA from heart, kidney, liver, and spleen of mature (10 to 12 months old) and aged (29 months old) mice were tested for their ability to translate viral RNA in a tRNA-dependent cell-free system.[3] However, the accumulation with age of enzyme molecules devoid of or with reduced catalytic activity is well established.[4]

It appears that the formation and accumulation of altered enzyme molecules is a universal phenomenon; altered forms of enzymes have been detected as a function of age in the nematode,[5-8] mouse,[9,10] rat,[11] rabbit,[12] and man.[12-13] For a variety of enzymes the active molecules of old animals cannot be distinguished from those of the young by electrophoretic mobility, affinity for the substrate, immunological activity, nor molecular size, but they do show noticeable differences in thermal stability.[5,6,9-11] Subtle changes in enzyme conformation or structure may take place in vivo with age while antigenic cross-reactivity with respect to the enzyme from the young organism remains complete.[1,14] It has been suggested that the proportion of inactive enzyme molecules encountered in old animals is potentially detrimental, and that a decline in efficiency of the degradation system in older animals may account for the occurrence of altered protein molecules in aging organisms.[7] Post-translational modification of protein molecules could also induce molecular polymorphism. Much of the protein hormone polymorphism occurring in normal animals, animals with endocrine disorders, and perhaps in aged animals probably involves post-translational changes of various types, some in the secretory cells, others in the circulation.

Many polypeptide hormones occur in several different forms even in the normal animal. These polymorphs, or heterogeneous forms, occur because the biosynthetic and catabolic mechanisms involved may not be precisely defined, thus allowing "leakage" of a small percentage of changed molecules. In the hurly-burly of the supply-transport-secretion cycles, incompletely synthesized forms may be released or inappropriately stored. Certainly, under endocrine demand or pathological crisis polymorphs become more common.[15] At present only a handful of studies have discovered differences in hormonal forms associated with aging.[16-20]

The purpose of this chapter is to present a survey of polypeptide hormone polymorphism, particularly the anterior pituitary hormones. Kinds of polymorphism will be considered, as will some correlations with physiological studies. Relevance of hormonal polymorphism to the phenomenon of aging will be indicated. An attempt will be made to illustrate the discussion using data collected from a recent study into thyrotropin heterogeneity.[20]

II. POLYPEPTIDE HORMONE POLYMORPHISM

A. Categories of Polymorphism

A survey of the physicochemical characteristics of the anterior pituitary hormones shows that these hormones can be found in a multitude of molecular forms (Tables 1

Table 1
HETEROGENEOUS FORMS OF ANTERIOR PITUITARY HORMONES

	Heterogeneity		Source of hormone		
Hormone	Type	Number	Tissue	Species	Ref.
Andrenocortico-tropic hormone (ACTH)	Size	2	Pituitary, plasma	Man, rat	21, 22
	Size	2	Cellfree synthesis	Mouse	23
	Size	3	Pituitary cells in culture	Mouse	24
	Size	4	Pituitary	Mouse, rat	25, 26
	Size	5	Hypothalamus	Rat	26
Follicle stimulating hormone (FSH)	Size	2	Pituitary	Man	27
	Size, MCR	2	Pituitary	Monkey	28
	MCR	Several	Pituitary	Rat	29
	CC	2	Pituitary	Hamster	30
Growth hormone (GH)	Size	2	Pituitary, pituitary perfusion	Man, rat	31, 32
	Size	4	Serum & pituitary incubation	Man	33
	Size, MCR	4	Pituitary	Man	34
	Size	5	Pituitary	Man	35
	IR	Several	Pituitary	Pig	36
Luteinizing Hormone (LH)	Size	2	Pituitary	Man	37
	Size, MCR, CC	2	Pituitary, serum	Monkey, rat	19, 28, 38
	Size, BA, CC	2	Pituitary incubation	Rat	39
	IR, BA	Several	Plasma	Man	40
	IR, RRA	Several	Serum	Monkey	41
	IR, EPM	6	Pituitary, serum	Rat	42
LH beta-subunit	Size, CC	2	Cellfree synthesis	Cow	43
Prolactin (PrL)	Size	2	Serum	Cow, goat	44
	Size	2	Pituitary, plasma, serum	Man, mouse	15, 33, 45, 46
	Size	3	Pituitary, pituitary culture, plasma	Rat	47, 48, 49
	IR, BA	Several	Serum	Rat	50

Note: Abbreviations: BA, biological activity; CC, carbohydrate composition; EPM, electrophoretic mobility; IR, immunological reactivity; MCR, metabolic clearance rate; RRA, receptor binding activity; size, molecular size.

and 2). In some contexts, the spectrum of protein forms has made the identification of any one of these hormones by a single term quite inadequate;[65] five forms of growth hormone (GH)[35] have been described, six forms of luteinizing hormone (LH),[42] and five forms of thyroid stimulating hormone (TSH, thyrotropin).[57] It is, therefore, quite evident that hormone heterogeneity exists in the circulation and/or in the site of synthesis. The cause and physiological significance of hormone heterogeneity are still a subject for speculation and experimental scrutiny.

The identification of a hormone fraction with a molecular size larger or smaller than purified hormone may suggest the existence of: (1) precursors, representing various stages of biosynthesis; (2) noncovalently bound dimers or polymers; (3) hormone-binding proteins; (4) artifacts, perhaps the result of the extraction or the method of analysis; or, (5) hormonal subunits and fragments. In general, size heterogeneity can be investigated using filtration or chromatography procedures that can separate a molecular mixture according to size. This is probably the simplest and most economical kind of study to undertake, especially if an assay of high sensitivity and specificity is available to measure the levels of hormones in the fractionated sample. Unfortunately, this method is likely to reveal a minimum number of different forms because of the limited differentiating power of ordinary sizing columns; Tables 1 and 2 show that the number

Table 2
HETEROGENEOUS FORMS OF THYROID STIMULATING HORMONE (TSH) AND TSH SUBUNITS

Hormone	Heterogeneity		Source of hormone		Ref.
	Type	Number	Tissue	Species	
Thyroid stimulating hormone (TSH)	Size	2	Pituitary, plasma	Man, rat	17, 18, 51, 52, 53
	Size, IR, RRA	Several	Pituitary tumor, serum	Mouse	54
	IR, CBA	Several	Serum	Man	55
	EPM	4, 5	Pituitary	Man, rat	56, 57
	EPM, CC	4	Pituitary	Whale	58
	EPM, BA, CC	4	Pituitary	Cow	59
TSH alpha-subunit	Size	2	Pituitary, serum	Man, rat	53, 60
	EPM	4	Pituitary	Man	61
TSH beta-subunit	Size	2	Pituitary, serum	Man, rat	53, 62, 63
	EPM	4	Pituitary	Man	61
	ASA	2	Pituitary	Cow, man	64

Note: Abbreviations: See Table 1; ASA, affinity for alpha-subunit to form TSH; CBA, potency in cytochemical bioassay.

of size polymorphs is generally less than that of other forms. Nevertheless, the presence of "big" or "small" hormones can yield considerable information about biosynthetic pathways, and can be indicative of some types of post-translational modification. In any initial study it is advisable to look for size heterogeneity before more powerful methods of analysis are tried.

Other types of protein hormone heterogeneity are independent of molecular size. This kind of heterogeneity, termed microheterogeneity, may affect molecular charge and probably reflects variations in carbohydrate composition, amino acid deamination, or single amino acid substitution (Tables 1 and 2). Microheterogeneity can be readily identified since the electrophoretic mobility of the molecule may be drastically altered without a change in molecular size; the number of polymorphs separated by electrophoresis is, therefore, greater than the number of forms separated on a gel filtration column. If carbohydrate composition of glycoprotein hormones (LH, FSH, TSH, hCG, and their respective subunits) is drastically altered then there will also be a difference in molecular size. Carbohydrate composition, or degree of glycosylation, may also affect the biological and immunological identity of the hormone so that most of the kinds of polymorphism for the glycoprotein listed in Tables 1 and 2 could be attributed to changes in glycosylation. A further category of size-independent heterogeneity relates the molecular "shape" or "conformation". Generally, "conformational heterogeneity" affects biological and/or immunological activity[36] so that an immunological approach can be used for its characterization. Such studies involve the use of two radioimmunoassays, one which recognizes an antigenic determinant dependent on conformational integrity. The second radioimmunoassay recognizes an antigenic determinant that is not accessible to the assay antibody unless there is some degree of hormonal unfolding.

It can be readily recognized then, that microheterogeneity and conformational heterogeneity should be studied with at least two different kinds of analytical tool. For example, a comparison of immunological and biological activities may reveal subtle changes not evident when only one method is used. This is because the sites of immunological identity and biological identity are not usually the same. In that case, alteration of one or other site can be seen as an alteration in the ratio of immunological to biological potencies. Several examples are listed in Table 1.

B. Glycoprotein Hormone Polymorphism

The physiological significance of hormone polymorphism is not yet understood, but it seems clear that the number of different hormonal forms can be affected by the physiological state of the organism.[15] Different forms may be secreted independently in some circumstances. However, it should be noted that the appearance of very high-molecular weight forms of prolactin in plasma and pituitary extracts may often be due to experimental manipulations (such as freezing and thawing) rather than to physiological causes.[49] Much of our present understanding of the relevance to physiological mechanisms of hormone polymorphism has come from studies of the pituitary glycoprotein hormones and human chorionic gonadotropin (hCG).

The glycoprotein hormones LH, TSH, FSH, and hCG consist of two chains or subunits held together noncovalently.[66] The two subunits, designated α and β, have little biological activity as individual free subunits.[67] The α-subunits of all four human glycoproteins are nearly identical,[68] but differ significantly in their carbohydrate moieties.[66,69] The amino acid sequences of the β-subunits differ and account for the unique immunological and biological activities of each hormone.[66,68,69] It is thought that the α- and β-subunits are synthesized separately from separate mRNAs, rather than together on a large precursor form.[39,70-74] After cleavage of the 24 amino acid signal peptide, the nascent subunits are initially core glycosylated by attachment of complex oligosaccharides containing mannose, glucose, and *N*-acetylglucosamine to specific amino acid residues. Excess mannose and glucose residues are trimmed from the intermediates, which are later modified by the sequential addition of terminal carbohydrate moieties (*N*-acetylglucosamine, galactose, *N*-acetylneuraminic acid, sialic acid) to complete the oligosaccharide structures. Disulfide bridges are completed before the mature α- and β-subunits are paired into a stable complex in a reaction that involves some conformational changes to achieve maximum stability. Some of the steps in the biosynthetic pathway are not completely understood, and indeed, the exact sequence is unknown.[71] Certainly the subject requires further study since the discovery of some of the regulatory aspects of glycosylation may explain the relationship between the carbohydrate composition and the biological properties of these hormones.

It has been demonstrated that the physiological state of the animal can influence the biochemical nature of pituitary LH and FSH. Studies in ovariectomized and estrogen-treated monkeys indicate that steroids may alter the content of sialic acid in pituitary LH.[38] Similarly testosterone and estrogen alter pituitary LH isolated from male and female rats.[29,75] The endogenous FSH circulating in the blood of androgen-treated rats differs from that in the blood of androgen-deprived rats. Also immunoreactive-FSH (IR-FSH) present in pituitary extracts differs from IR-FSH present in serum.[29] The qualitative changes of LH and FSH are characterized by a larger apparent molecular size and slower rate of disappearance from the circulation of test rats.[28] It is suspected that both the differences between pituitary and serum FSH, and the steroid-directed transmutations among stored and circulating forms of the glycoprotein, may involve differences in the composition of its carbohydrate side chains.[28,29]

Several recent studies indicate the importance of the carbohydrate side chains to the biological properties of glycoprotein hormones. Rat pituitaries, in the presence of gonadotropin releasing hormone, synthesize two kinds of IR-LH.[39] The high-molecular weight form seems to appear early in the biosynthetic pathway, to contain relatively little carbohydrate, and to possess little bioactivity. The other form is native LH. Kourides and co-workers have observed that circulating free α-subunit is larger in apparent molecular weight than intrapituitary α-subunit.[60] The intrapituitary α-subunit contains little galactose and sialic acid, thus accounting for the differences in molecular weight. Hence, the smaller forms of α-subunit are precursors of the larger forms. Even though glycosylation does not appear to be required for the secretion of glycopeptides,[72,76] terminal glycosylation of α-subunit with galactose and sialic acid is temporarily closely associated with the secretory process.[60,74] Specific glycosylation may be required for subunit combination,[72] and deglycosylation may cause inhibition of biologic activity.[77,78]

In summary, the cellular glycosylation of proteins is a dynamic process in which there is continued removal and addition of sugar residues during the transport of the glycoproteins through the secretory pathway. Glycosylation, while being important in the expression of hormonal activity, may not be necessary for secretion. Therefore, forms of a hormone may be secreted and recognized by immunoassay, but may not possess bioactivity.[79]

III. ALTERED POLYPEPTIDE HORMONES AND AGING

Biological theories of aging have frequently directed attention toward the brain and endocrines as control systems on the assumption that age-dependent alterations in these integrative systems must have widespread consequences throughout the body.[80] Hormones have a critical role in bridging the gap between molecular, cellular, and whole-organ levels of physiological organization. Therefore, hormones from the anterior pituitary, which is a major determinant of neuroendocrine activity, should be examined for their possible importance in the genesis of senescent deterioration.[81,82] Measurement of circulating pituitary and target gland hormone concentrations in man has not revealed any well-defined impairment of pituitary tropic hormone reserve with age. With the possible exception of GH, pituitary hormone secretion continues at the same, or at even higher levels for LH and FSH, throughout old age.[83] However, almost all hormones secreted by the anterior pituitary have been shown to influence aging phenomena, and furthermore, it has been suggested that the pituitary regulates the rate of aging and the onset of age-associated pathology.[81] On the other hand, hypophysectomy is known to retard aging processes and to inhibit the development of certain diseases of old age.[84] Hypophysectomy can return some structural age-related changes in the kidney to the values seen in younger animals,[85] and removal of the pituitary gland followed by thyroxine administration in old rats can lead to a return on a variety of biochemical and physiological parameters to levels found in young animals.[86]

If the pituitary hormones undergo qualitative changes such that their biological potency is significantly affected, it may be possible to reconcile the idea that the anterior pituitary plays a role in the basic mechanism of aging with the observation that the blood levels and pituitary stores of pituitary hormones are not consistently age-dependent. As already discussed above, certain post-translational modifications may severely inhibit the biological activity of hormones without affecting their immunological properties as detected by radioimmunoassay. Glycoprotein hormones are particularly dependent on their degree of glycosylation for their expression of biological activity. Perhaps some knowledge of hormone polymorphism may lead to a better understanding of aging mechanisms.

Klug and Adelman have found that a high-molecular weight form of IR-TSH (immunoreactive TSH) accumulates in the serum of old rats.[17,18] In 24-month-old rats, up to 66% IR-TSH occurred in the high-molecular weight region of a gel filtration elution profile, compared to only 12% in 2-month-old animals.[18] The pituitary also contained some high-molecular weight IR-TSH but not in such large percentages. According to a McKenzie bioassay,[87] the large form of IR-TSH had thyroid-inhibiting activity, but immunological activity, as measured in a radioimmunoassay, was unimpaired. The occurrence of TSH heterogeneity in aged rats has recently been confirmed,[20] and further data will be presented below. LH in the pituitary and circulation of old male rats (24 months old) appears to have a larger molecular size than normal.[19] Neuraminidase treatment of pituitary extracts abolished the molecular size difference, thus suggesting that the increased size was due to an increase in sialic acid content of LH. This was verified by a study of metabolic clearance kinetics of the large LH injected into young rats. The larger form of LH had a slower metabolic clearance rate; this data was analogous to the results reported earlier by Peckham and Knobil for studies on the ovariectomized and estrogen-treated monkey.[38] Thus, testosterone administration to old rats produced a normalization in molecular size, and castration of young rats caused an increase in size.[19] No other data is yet available concerning the age-associated polymorphism of anterior pituitary hormones. Only one other example of the phenomenon of age-related hormone heterogeneity can be gleaned from the current literature. Duckworth and Kitabchi have reported that after glucose challenge older subjects have a greater concentration of proinsulin-like material (PLM) in the circulation.[16] The amount of PLM showed a significant correlation with the age of the subject, with the overall difference being approximately twofold. The only criterion for supposing PLM was really the high-molecular precursor of insulin, proinsulin, has been the proinsulin radioimmunoassay, so these observations undoubtedly require further confirmation. It is possible to suppose that the increased circulating PLM represent inappropriate release of "unfinished" product from the pancreas, but it is also possible that such accumulation represents an altered rate of peripheral degradation since patients with chronic renal disease also show increased PLM levels.[88]

IV. ANALYSIS FOR TSH POLYMORPHISM IN AGED RATS

The remainder of this chapter will cover the theoretical and methodological considerations involved in a study of TSH polymorphism in old rats. General discussion will include choice of animals, assays, and fractionation procedures; none of these factors are considered optimum as can be seen from the resultant data.

A. Animals

A major problem for many gerontological investigators seems to be finding a source of old rats. It is possible to maintain a colony of animals locally, but this can be expensive and very time-consuming since the life span of most strains of rats is up to 3 years. It is also probable that different rat colonies, of the same strain but maintained in separate laboratories, will have different age-associated pathology and mean life span. Duplication of experimental findings can therefore be difficult. Such objections need not be serious since aging phenomena might be expected to be qualitatively, if not quantitatively, similar. It is possible to purchase 12-month-old rats, usually retired breeders, from commercial breeders. If these animals are maintained for a further 12 to 18 months in the laboratory, a viable alternative that is economical in time and cost is available. However, this source of old rats is not universally accepted by investigators, especially when certain neural and endocrine systems are to be studied.

The best source of old rats is a stock colony specifically raised and kept for gerontological study. Such a colony would be from a caesarian-derived nucleus stock and maintained under barrier husbandry conditions. For example, the rearing, maintenance and age-related pathology of a Sprague-Dawley strain of rats has been described by Cohen and co-workers.[89] Old rats of the Fischer 344 strain may be obtained from the National Institute on Aging (NIA). This colony is barrier-maintained at the Charles River Breeding Laboratories (CRBL), Wilmington, WA. NIA approval is required before animals can be bought from the CRBL colony. Rats from these sources are assuredly of the correct age and would be expected to be in optimum condition. How-

Table 3
PITUITARY TSH LEVELS IN RATS OF VARIOUS AGES

	Pituitary			Concanavalin A-column yield (μg/gland)[a]		
Age (mo)	TSH content (μg/gland)	Gland weight (mg)	TSH concentration (μg/mg tissue)	G Fraction	NG Fraction	Total G + NG
3	640 ± 230(5)[b]	8.1 ± 0.2(5)[b]	79.5 ± 13.2(5)[b]	393[c]	21[c]	414
12	845 ± 188(5)	10.5 ± 1.2(5)	81.9 ± 7.5(5)	209	15	224
22	950 ± 327(5)	14.4 ± 0.6(5)	67.0 ± 11.8(5)	299	12	311
30	818 ± 266(4)	15.1 ± 1.6(4)	54.7 ± 7.2(4)	214	15	229

[a] Values for the concanavalin A-column were normalized with respect to a whole gland sample.
[b] Expressed as mean ± SEM. Number in () represents number of animals.
[c] Single determination.

ever, unit price can be prohibitively expensive. Several investigators with diverse research objectives may find it amenable to share resources by collecting only those organs of interest to each.

The chief recommendation for using animals from a central supplier is provided by the great body of experimental data already available. Certainly some knowledge of age-associated pathology in senescent rats is important since the incidence of neoplasms and lesions of various severity may influence the whole animal. The aged Fischer 344 rat exhibits a wide variety of pathology.[90-92] In particular, renal disease severity is highly correlated with increasing age.[92] Renal disease could be a possible cause for altered serum hormone profiles since excretion rates of hormones and metabolic fragments could be affected.

Few tumors are seen in Fischer 344 rats younger than 18 months. Testicular interstitial cell tumors, pituitary chromophobe adenomas, and a form of mononuclear cell leukemia are the most common types of tumor in older rats. For the study of pituitary hormones, the occurrence of pituitary adenomas are of some concern. It should be noted that pituitary hormones are produced by basophils and acidophils and may be unaffected by the chromophobe adenomas. The overall incidence of pituitary tumors is relatively low (up to 15%).[90,92]

The occurrence of a pituitary tumor need not necessarily exclude an animal from study. It could even be said that such a preselection would place undue weight on a healthy, tumor-free, part of the total population. Nevertheless, serum levels of pituitary hormones should be measured to identify animals having abnormal pituitary function. In our experience with the Fischer 344 we noted testicular tumors in rats older than 22 months. Pituitaries of old rats (22 and 30 months old) were enlarged compared to young adults (Table 3), but serum and pituitary TSH concentrations were not elevated (Table 4).

In the study described below, Fischer 344 rats (3, 12, 22, and 30 months old) from the NIA colony were used. Two to three weeks after delivery to our laboratory the rats were decapitated, and serum from trunk blood was collected and frozen in 0.5 mℓ aliquots at −70°C until required for analysis. Pituitary glands were stored at −70°C for later hormone extraction. Various organs were collected for histology, and brains were separated into various regions for monoamine analysis. These latter studies are not relevant to the present study but are mentioned to indicate how maximum use can be made of the animals.

B. Assays

1. Preliminary Considerations

The idea of assay-specific hormone polymorphism has been considered. In essence, assays rely on the detection of specific regions of a molecule; these regions need not

Table 4
SERUM TSH LEVELS IN RATS OF VARIOUS AGES

Age (mo)	Serum TSH concentration (ng/mℓ)	Sephadex column yield (ng/mℓ serum)[a] G fraction	NG fraction	Total G + NG
3	252.8 ± 35.2(5)[b]	53[c]	208[c]	261
12	340.8 ± 70.6(5)	70	290	360
22	228.6 ± 32.8(5)	22	159	181
30	103.1 ± 27.3(4)[d]	28	113	141

[a] Values for Sephadex column normalized with respect to unit volume serum sample previously fractionated on concanavalin A-column.
[b] Expressed as mean ± SEM. Number in () represents number of animals.
[c] Single determination.
[d] Significantly lower than 3-month-old rats, $p < 0.01$.

be the same for all assays. Other molecule entities, which are either entirely unrelated to the hormone or are cleaved fragments of the hormone, will be recognized by the assay if the molecules possess the key amino acid sequence(s). The question to be addressed must then concern the choice of assay type to be used to determine "hormone" concentrations. Any choice implicitly suggests which part of the hormone is to be accepted as being "most characteristic". Should the concentration of the hormone in a sample be considered more important than the specific biological activity of that hormone? No assay currently available can be considered ideal since they all possess either theoretical or practical limitations; the application of several different assays for one hormone can yield considerably more information (Tables 1 and 2).

Before the advent of radioimmunoassays, it was not possible to accurately measure the minute amounts of hormones in the systemic circulation. The limited number of bioassays (BA) of hormones in endocrine glands and biological fluids lacked both precision and sensitivity. Thyrotropin (TSH) is still assayed using various modified bioassays when it is necessary to verify that tissue extracts of TSH retain in vivo potency. The McKenzie bioassay measures release of isotopically labeled thyroid hormones in intact mice.[82,93,94] A more recent in vitro bioassay measures TSH-induced decrease in the concentrating capacity for unbound iodide of guinea pig thyroid tissue.[95] A further assay, which can be described as a bioassay because of its fundamental physiological basis, is the cytochemical bioassay (CBA).[96] The CBA may eventually have the precision, sensitivity and convenience of ligand-binding assays. TSH induces endocytosis of colloid by the thyroid follicle cells, and the fusion of the endocytotic vesicles with lysosomes. The response to TSH is reflected by a change in the stability of the lysosomal membranes. In the CBA, membrane stability determines the rate at which an added chromogenic substrate for lysosomal enzymes penetrates the lysosomal membranes. The color reaction may be measured by microdensitometry in single cells of thyroid tissue incubated with the test medium and appropriately prepared for microscopy. In the present stage of development the CBA lacks speed, and the number of samples that can be measured is severely limited. However, in a recent study, sera from two hypothyroid patients with pituitary-hypothalamic disease were analyzed for TSH by radioimmunoassay (RIA) and CBA. TSH concentrations measured by RIA were elevated, but the CBA showed a normal value. It was suggested that the pituitary gland of these patients secreted a biologically inactive form of TSH.[55] Undoubtedly the CBA has great potential for endocrinology since the techniques may be modified and applied in the assay of many other hormones. Until then the assays of primary choice are the competitive binding assays.

Competitive binding assays have their basis in the competitive inhibition by unlabeled hormone to its specific antibody (radioimmunoassay, RIA) or membrane receptor (radioreceptor assay, RRA).[97] The RIA measures an antibody-antigen interaction; the chief objection to this assay is that it determines immunogenic identity based on a part of the molecule remote from the site of biological identity. The assay can recognize molecules that have no biological activity. For example, in the measurement of prolactin concentrations, BA and RIA may be highly correlated, but the two assays measure different proportions of the "prolactin" in the sample. It has been reported that RIA measures only about 25% of the hormone that is detected by mammary gland organ BA.[50] If some knowledge of structure-function relationships is available for the hormone being studied it may be possible to obtain a better correlation between BA and RIA. Since the biological identity of the pituitary glycoprotein hormones resides in the β-subunit, an assay for the β-subunit only, rather than for the whole molecule, may give a truer value for the concentration of active hormone. It should not be assumed, however, that the RIA necessarily measures more hormone than the BA. When measurement of pituitary TSH concentrations were performed in patients with asymptomatic atrophic thyroiditis it was shown that bioassayable TSH was elevated, but the radioimmunoassay failed to confirm any difference.[98] In this example, some of the TSH reacted in the BA but had only a weak affinity for TSH antibodies.

Unlike the RIA, the radioreceptor assay (RRA) does have a more justifiable claim as a BA-like procedure. The binding agent is a TSH-binding membrane receptor isolated from thyroid tissue so that the binding reaction partly emulates the action of the TSH in vivo.[99,100] Since membrane binding does not necessarily imply a cellular response the RRA cannot be called a bioassay. It can be considered as an intermediate between BA and RIA; heterogeneous forms of TSH in mouse thyropic tumor and serum differ in RIA and RRA activities.[54]

2. TSH Radioimmunoassay

Reagents for the TSH RIA are available from Dr A. F. Parlow of the NIAMDD Rat Pituitary Hormone Distribution Program. The procedures for preparing labeled TSH, and for processing the assay are described in protocol sheets supplied with the reagents. The basic methodology used in the RIA procedure has been described and can be considered to be uncomplicated.[101] Several modifications can be made to overcome the limiting sensitivity of the TSH radioimmunoassay; a high sensitivity is required to assay samples from a fractionation column so that a complete elution profile can be obtained. A five- to tenfold increase in sensitivity can be readily achieved by adding labeled hormone 24 hr after beginning a preincubation of the antibody, sample, and buffer.

In brief, the procedure is as follows. Into 10 × 75 mm plastic test tubes are added: 20 $\mu\ell$ assay buffer (0.01 *M* phosphate buffer pH 7.6; 0.15 *M* NaCl; 0.01% sodium azide; 1% bovine serum albumin), 300 $\mu\ell$ sample or TSH reference standard, and 200 $\mu\ell$ antibody solution (1:10 000 dilution; 0.01 *M* phosphate buffer; 0.15 *M* NaCl; 0.05 *M* EDTA; 3% normal rabbit serum; 0.01% sodium azide). The solution is vortexed briefly and incubated at room temperature for 24 hr. Radioiodinated TSH (100 $\mu\ell$; approximately 20,000 cpm activity; 0.1% bovine serum in 0.01 *M* phosphate-buffered saline) is added, and the incubation is allowed to continue for a further 24 hr. The second antibody, sheep or goat antirabbit gamma globulin is finally added (200 $\mu\ell$; 1:10-1:20 dilution in 0.01 *M* phosphate-buffered saline) for the last 24-hr incubation. The precipitated antibody-antigen complex is separated by centrifugation (1000 g for 30 min) and aspiration of the supernatant. The precipitated radioactivity is measured and results finally expressed as ng NIAMDD-rat TSH-RP-1 per unit volume. Provided that the second incubation (after addition of label) is not prolonged, the standard curve for the assay is in the range 2.5 to 100 ng/tube compared to 19 to 300 ng/tube when there is no preincubation. The shorter assay is adequate when samples from pituitary glands are to be analyzed, but in column fractions of serum derived extracts the more sensitive assay is required.

Reagents for TSH beta-subunit can be obtained from the NIAMDD Rat Pituitary Hormone Distribution Program. Once again, the procedures to follow are outlined in information sheets sent with the reagents. In general, it is possible to use the same methods as described for TSH. The assay can be used to measure pituitary concentrations of the polypeptide subunit, but is more difficult to measure levels in serum and in column fractions. Cross-reactivity between the two assays can be determined by measuring the amount of added TSH in a sample with the TSH-beta assay, and vice versa. In general, TSH has a very small amount of activity in the TSH-beta assay (0.04% cross-reactivity) whereas TSH-beta is readily recognized by the TSH assay (complete cross-reactivity; 1 ng TSH-beta equivalent to 90 ng TSH). Therefore, TSH values may be incorrect if significant amounts of TSH-beta are also present in the sample. On the other hand, unless TSH is present in large quantities compared with TSH-beta, the TSH-beta assay should accurately reflect TSH-beta concentrations.

C. Gel Filtration Chromatography

Suitable ultrafiltration media for the separation of protein mixtures according to size can be obtained from Pharmacia or Biorad. The grade and molecular weight range used is determined by the sample components. TSH samples are chromatographed on a 1.8 × 90 cm column of Sephadex® G-200 (Pharmacia) equilibrated in phosphate-buffered saline (0.05 *M* phosphate buffer, pH 7.4; 0.8% NaCl; 0.1% bovine serum albumin). Flow rate, controlled with a suitable peristaltic pump, is 9 mℓ/hr, and 70 3-mℓ-fractions are collected. All runs are at 4°C. Column fractions are either immediately assayed for TSH, or stored at −20°C for later analysis. Pituitary-derived fractions may be assayed directly because of their high hormone concentrations. Serum-derived fractions are lyophilized and reconstituted in 0.5 mℓ water to concentrate the protein by a factor of six prior to RIA. Results can be expressed as ng TSH/mℓ serum, ng TSH/fraction, or ng TSH/fraction/mℓ serum; this depends on whether the sample applied to the gel filtration column is serum, a pituitary extract, or a serum extract. Normalization of data with respect to unit volume serum where appropriate allows a comparison of peak size from different column runs with different sample sizes. To further facilitate comparison between experiments, the Sephadex column is calibrated by separate runs with Dextran blue, [^{125}I]-labeled TSH, and [^{125}I]-iodide, and also by adding tracer amounts of [^{125}I]-TSH and [^{125}I]-iodide to some of the samples. The positions of these markers are aligned to allow correct comparison of the various molecular weight profiles of TSH.

The proportion of heterogeneous forms in the elution profile is determined by dividing the profiles into three sections. The position of these cuts is taken from the peak position in a molecular weight profile of samples taken from young adult animals (see Figure 5). The amount of immunoreactive TSH (IR-TSH) within each area is then expressed as a percentage of the total amount of IR-TSH recovered from the column. It is then possible to compare the percentages across the age groups.

It is possible to gain a qualitative appreciation of the occurrence of thyrotropin heterogeneity from a simple fractionation of serum on a gel filtration column. Data from such an experiment are illustrated in Figure 1. Sera (3.0 mℓ) from a 6-month-old (A), a 12-month-old (B), and a 27-month-old (C) Long-Evans rat were applied to the Sephadex column and the fractions were assayed immediately by TSH RIA. Results were expressed as ng TSH/mℓ of fraction. These data suggested an apparent increase in high-molecular weight IR-TSH with increasing age and confirmed previously re-

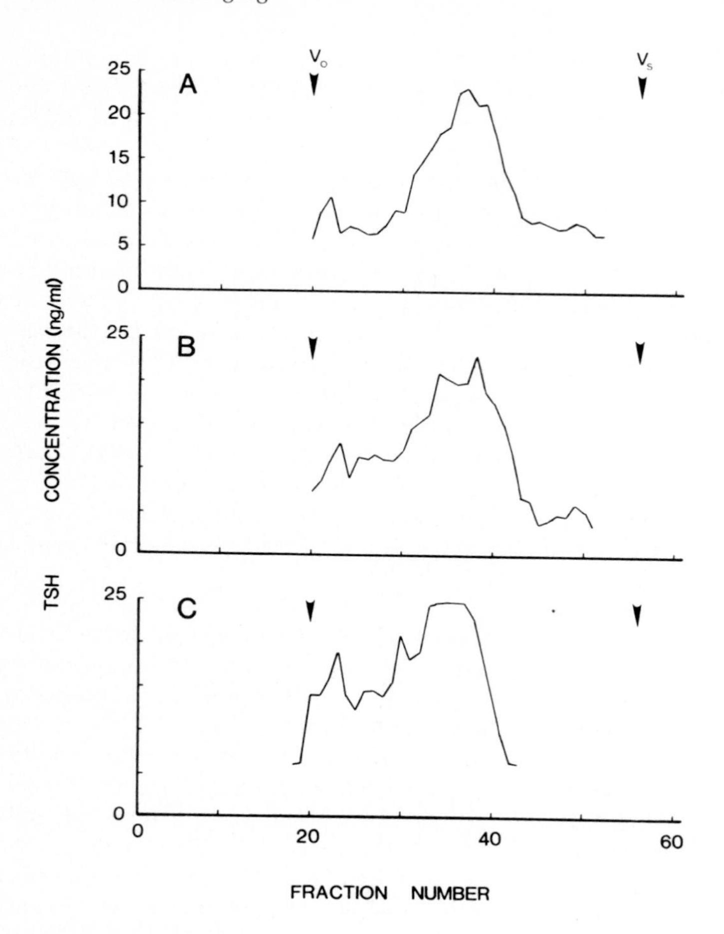

FIGURE 1. Sephadex® G-200 fractionation of serum from rats aged 6 months (A), 12 months (B), and 27 months (C). Arrowheads mark the position of the column void volume and salt volume, V_o and V_s, respectively.

ported studies.[17,18] However, this information does not indicate the nature or possible causes of the increased TSH heterogeneity. Further analysis, other than gel filtration fractionation, is required to yield better characterization of the phenomenon. Apart from molecular size, the next most obvious physicochemical property to investigate is carbohydrate composition.

D. Affinity Chromatography of Glycoproteins

1. Introduction

Lectins or phytohemagglutinins extracted from certain plants are specific adsorbents of carbohydrates.[102] One of these lectins, the crystalline Jack bean globulin, concanavalin A, is known to interact with the carbohydrate moieties of serum glycoproteins.[103] Hence, concanavalin A immobilized on a neutral support medium, like agarose beads, has been used for the affinity chromatography fractionation of glycoprotein mixtures,[104] and the technique seems desirable in the purification or extraction of glycoprotein hormones from various tissues, plasma, and urine. The method can be a very efficient way for the separation of pituitary glycoprotein hormones.[104-106] In some instances it has been possible to extract hormones from samples of urine and plasma before RIA in order to overcome the sensitivity limits of the assay.[107]

Concanavalin A affinity chromatography could also be worthwhile in investigations of hormonal polymorphism.[30] This is especially so, in view of the relationship between the carbohydrate composition and the biological potency of glycoprotein hormones. The procedure could be applied together with RIA measurements to study immunoreactive glycosylated and nonglycosylated hormonal forms.[30,78]

2. Procedures

A detailed description of concanavalin A-Sepharose chromatography can be found elsewhere.[30,102,104-106]

Concanavalin A (Sigma) is coupled to Sepharose 4B (Pharmacia) using the method described by Dufau and co-workers.[104] First, the Sepharose (40 g suspended in 40 mℓ water) is activated by the addition of 8 g cyanogen bromide dissolved in 40 mℓ water. The cyanogen bromide solution is added slowly while the mixture is stirred continuously, and the pH maintained at 11 by addition of 4 *M* NaOH. The activated Sepharose is washed with 100 mℓ distilled water, then 100 mℓ 0.1 *M* $NaHCO_3$. Finally, the Sepharose is added quickly to a solution of 400 mg concanavalin A, 40 mℓ 1 *M* NaCl and 40 mℓ 1 *M* $NaHCO_3$. The reagents are mixed at 4°C for 18 hr, and the concanavalin A-Sepharose preparation is then washed with 2 l 0.1 *M* $HaHCO_3$ on a sintered glass filter. The preparation is stored in phosphate-buffered saline (pH 7.4) at 4°C.

Two sizes of column are used, a small column (1 × 5 cm) for samples of small volume (less than 0.5 mℓ), and a larger column (1.8 × 20 cm) for samples up to 8 mℓ volume. Both columns are packed with concanavalin A-Sepharose and the Sepharose is equilibrated with Dulbecco's phosphate-buffered saline (pH 7.4) (dPBS). After each run, the concanavalin A-Sepharose may be regenerated with several wash cycles of alternately acetate buffer (pH 4.6) and dPBS. Both columns are used in the same way, except that the flow rate of the small column is driven by gravity feed and adjusted with a stopcock, whereas the flow rate of the larger column is controlled with either a peristaltic pump or a constant-flow Marriot flask. The flow rate to be used must be determined by running standards (for example, ^{125}I-TSH) and calculating yields.

Hormone preparations (pituitary extracts to the small column, serum to the large column) are applied to the columns and elution is commenced with dPBS and continued for at least two column volumes. The elution buffer is then changed to dPBS containing 0.2 *M* methyl-α-D-glucopyranoside (Calbiochem) and continued for approximately four column volumes. Appropriate fractions are collected from the column. These may be assayed immediately for protein concentration[108] or for TSH concentration to yield an elution profile (see Figure 3). Fractions containing nonglycosylated material (eluted first) and glycosylated material (eluted by methyl-α-D-Glucopyranoside) are pooled separately, dialysed against 0.1 *M* ammonium bicarbonate, and freeze-dried. The two fractions, nonglycoslyated (NG) and glycosylated (G), are stored at −20°C until required for further analysis.

E. Scheme for Analysis of TSH Polymorphism

A suggested scheme for the investigation of TSH polymorphism in the blood and pituitary gland of aged rats is outlined in Figure 2. The series of experiments, using procedures described above, is designed to yield data concerning the carbohydrate composition and molecular weight profile of TSH polymorphs. Serum samples and pituitary extracts are treated in approximately the same manner.

Trunk blood is collected from Fischer 344 rats of various ages (3 months, 12 months, 22 months, and 30 months old). An equal volume of serum is taken from each rat within one age group to form a pool representative of that age group. This pool is separated on a concanavalin A-Sepharose column into unbound fractions and bound column fractions. The sets of fractions are pooled into a nonglycoprotein fraction

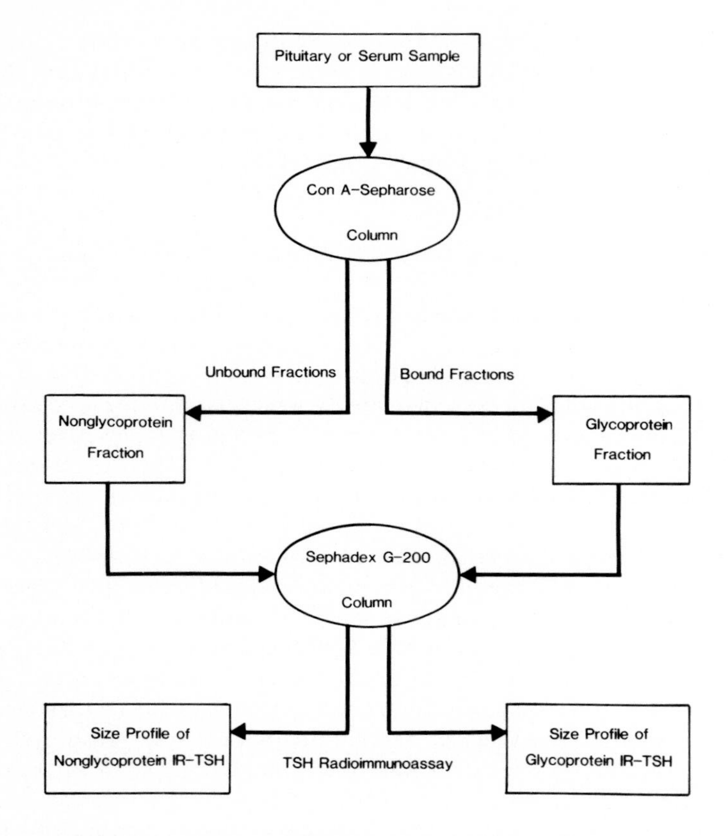

FIGURE 2. Scheme for the study of carbohydrate composition and size heterogeneity of immunoreactive TSH (IR-TSH) in serum and pituitary extracts from rats of various ages.

(NG) and a glycoprotein fraction (G), respectively. After concentration by dialysis and lyophilization, NG and G fractions are further chromatographed on the Sephadex® G-200 column. Each fraction (3.0 mℓ) from the Sephadex column is lyophilized, reconstituted in a small volume (0.5 mℓ) of water, and then assayed by TSH RIA to give the final molecular size elution profiles of IR-TSH. The elution profiles for NG and G can be compared, and if corresponding parts of the profile are summed, a derived molecular weight profile for whole serum (NG + G) can be obtained. Similarly, samples of pituitary homogenate from each rat are pooled and separated into NG and G fractions on the concanavalin A-Sepharose column. The unbound and bound fractions may be assayed for TSH to verify the positions of the NG and G peaks. The nonglycosylated material and glycosylated material may then be sized on the Sephadex column to obtain the respective molecular size elution profiles of IR-TSH (or TSH-beta subunit).

Molecular size elution profiles are divided into three regions corresponding to high-molecular weight IR-TSH, endogenous IR-TSH, and low-molecular weight IR-TSH. The amount of material in each region is expressed as a percentage of the total recovered IR-TSH. These percentages then represent the degree of polymorphism in the original samples, and can be compared for the four age groups.

Several possible problems should be highlighted when this type of experimental approach is to be used. First, the data are derived from pooled samples rather than from separate experiments on each rat. This is necessary in order to have sufficient sample for the complete analysis. Second, it is possible that experimental manipulations are a

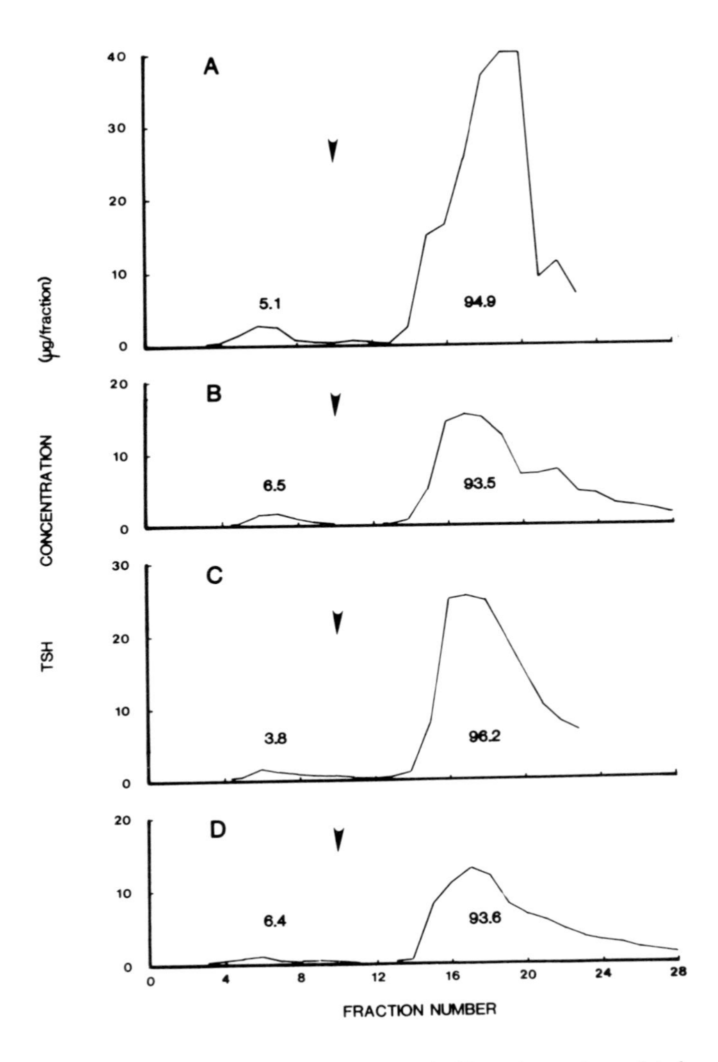

FIGURE 3. Separation of glycosylated (G) and nonglycosylated (NG) IR-TSH from the pituitaries of rats aged 3 months (A), 12 months (B), 22 months (C), and 30 months (D). The numbers on the graphs are percentages relative to the total IR-TSH recovered from the concanavalin A-column. Arrowheads mark buffer changes to elute bound IR-TSH.

source of molecular heterogeneity so it is important that all samples are treated identically. Other problems will be discussed with the data presented in the next section.

F. TSH Polymorphism in Aged Rats

Figure 3 illustrates the elution pattern obtained when pituitary extracts were chromatographed on a concanavalin A-Sepharose column. Approximately 3.8 to 6.5% of IR-TSH passed through the column as nonglycosylated material (as shown by the peak to the left of the arrowhead). This amount was similar to the 2 to 3% previously reported for bovine pituitary extracts.[106] The amount of nonglycoslated IR-TSH did not change appreciably with advancing age (A, 3 months old; B, 12 months old; C, 22 months old; D, 30 months old). These data, summed for the total TSH in each peak and then normalized with respect to pituitary gland content, was compared to TSH pituitary content determined from the original pituitary extracts (Table 3). Yield

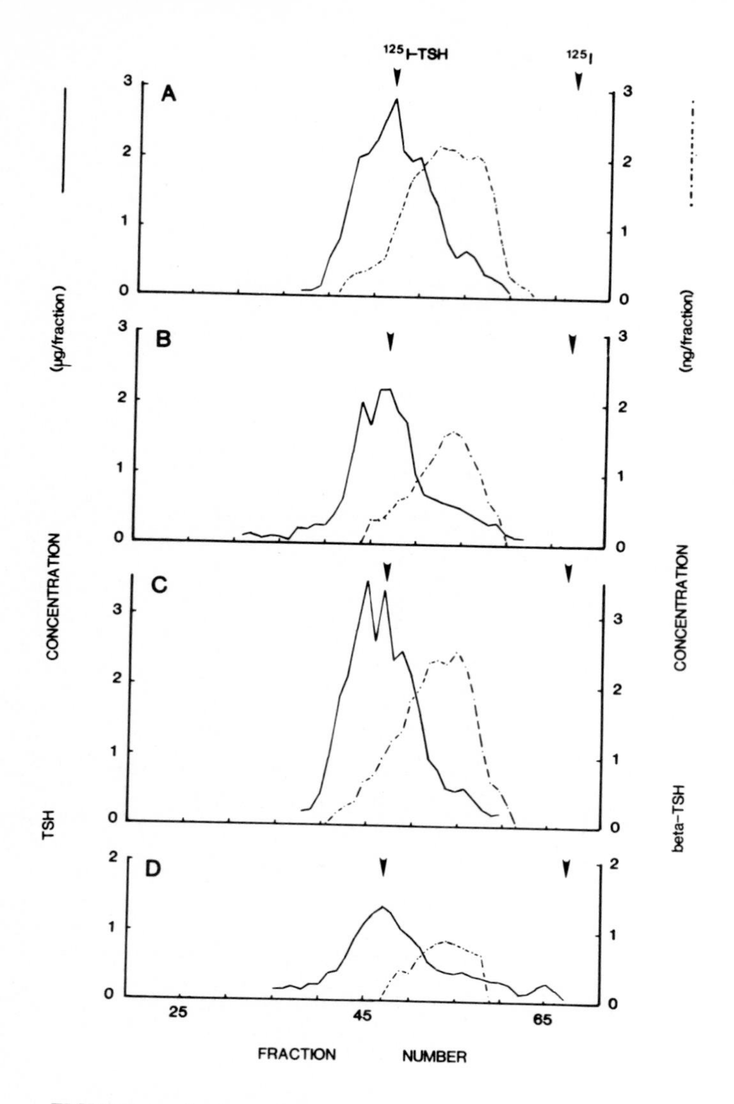

FIGURE 4. Sephadex® G-200 fractionation of glycoprotein pituitary extracts from rats aged 3 months (A), 12 months (B), 22 months (C), and 30 months (D). The pituitary extract was the G fraction from the concanavalin A-column. The black line represents IR-TSH, the dotted line, the beta-subunit of TSH. Arrowheads mark the relative elution position of ^{125}I-TSH, and ^{125}I-iodide.

from the column was approximately 30%; reasons for the loss are unclear. Data in Table 3 show that the TSH concentration in pituitaries of various ages decreased only slightly, and this change was not statistically significant. Total pituitary content did not change.

The eluted IR-TSH fraction comprising 95% of the total pituitary IR-TSH recovered from the concanavalin-A column (second peak in Figure 3) was rechromatographed on the Sephadex® G-200 column to give the molecular size profile of IR-TSH and IR-TSH-beta subunit for rats in four age groups (Figure 4: A, 3 months old; B, 12 months old; C, 22 months old; D, 30 months old). Figure 4 shows that IR-TSH and IR-TSH-beta were similar for all ages. Several details of the elution profiles were suggestive of differences in molecular size, but further studies using a method of chromatography with a better resolution are necessary to verify that proposal. The shape of the IR-TSH elution profile appeared to indicate the presence of several distinct species. On the low-molecular side of the main peak (marked by arrowhead corresponding to elution volume of ^{125}I-TSH calibration standard) there was a partially obscured minor peak corresponding to IR-TSH-beta cross-reactive with the TSH RIA. The presence of IR-TSH-beta was also detected with the TSH-beta RIA (dotted line). A third molecular species may have been detected, as indicated by the partial peak on the high-molecular side of the main peak. This became increasingly apparent with age (Figure 4, Panels A-C), but was not obvious in the oldest group (Panel D, 30 months old). The position of this material, which is only proposed from this data, relative to the principal IR-TSH peak was similar to the high-molecular weight LH found in the pituitaries of old rats.[19]

When serum was separated into glycosylated (G) and nonglygosylated (NG) fractions, and these separated fractions were rechromatographed on a Sephadex® G-200 column, molecular size elution profiles for IR-TSH were obtained (Figures 5 and 6). Data from Figures 5 and 6 were used to derive a profile for whole serum by adding together equivalent parts of the two curves (Figure 7). Furthermore, by summing the amount of IR-TSH in each column fraction it was possible to determine the recovery of IR-TSH from the Sephadex column. These data were used to calculate the percentage IR-TSH in various parts of the elution profile (Figures 5 to 7), and also to determine yields compared to serum IR-TSH concentration (Table 4).

Serum TSH levels decreased with increasing age (Table 4) with the oldest animals (30 months old) having a serum TSH concentration equal to 40% of the normal adult rat (3 months old). Data from the Sephadex chromatography showed the same trend for both G and NG fractions. Also, comparison of G + NG totals to serum levels (Table 4) indicated almost a complete recovery of IR-TSH despite the experimental manipulations required for the analysis (Figure 2). The amount of IR-TSH in the NG fraction exceeded that in the G fraction in all samples. This did not mean that 80% of the serum IR-TSH was nonglycoslated. Rather, these data show that the concanavalin A-Sepharose chromatography extracted from the serum a glycosylated fraction of IR-TSH and left poorly glycosylated or nonglycosylated IR-TSH.

The molecular weight profiles of glycosylated IR-TSH (Figure 5) showed an increase in hormone polymorphism with advancing age; an increased proportion in both high- and low-molecular weight IR-TSH species was apparent, with an increase of up to three to fourfold in 30-month-old rats. Nonglycoslyated IR-ISH showed similar increases in molecular size-heterogeneity (Figure 6).

The derived elution patterns for whole serum (Figure 7) were simlar to those obtained when whole serum was applied to the Sephadex column (Figure 1). At 3 months or 6 months of age (Figure 7, panel A; and Figure 1, panel A), the profiles had well-defined IR-TSH peaks. In 12-month-old rats (Figure 7, panel B; Figure 1, panel B) some size heterogeneity was already detected, as was seen from the changed peak separation. The trend continued in older rats (Figure 7, panel C, 22 months old; Figure 1, panel C, 27 months old). Approximately 66% of circulating IR-TSH in 24-month-old Sprague-Dawley rats appeared in the high-molecular weight region of Sephadex® G-200 chromatography elution patterns.[18] The corresponding amount was 53.6% in the 22-month-old Fischer 344 rat (Figure 7, panel C). It is proposed that most of the high-molecular weight IR-TSH is poorly or nonglycosylated, since after concanavalin A affinity chromatography, only 15.4% of circulating glycosylated IR-TSH is high-molecular weight material (Figure 5, panel C).

In summary, data presented in this section indicate that IR-TSH in the circulation and pituitary gland becomes more heterogeneous with advancing age. Pituitary IR-TSH appeared to shift to a higher molecular size, but the evidence for this was not

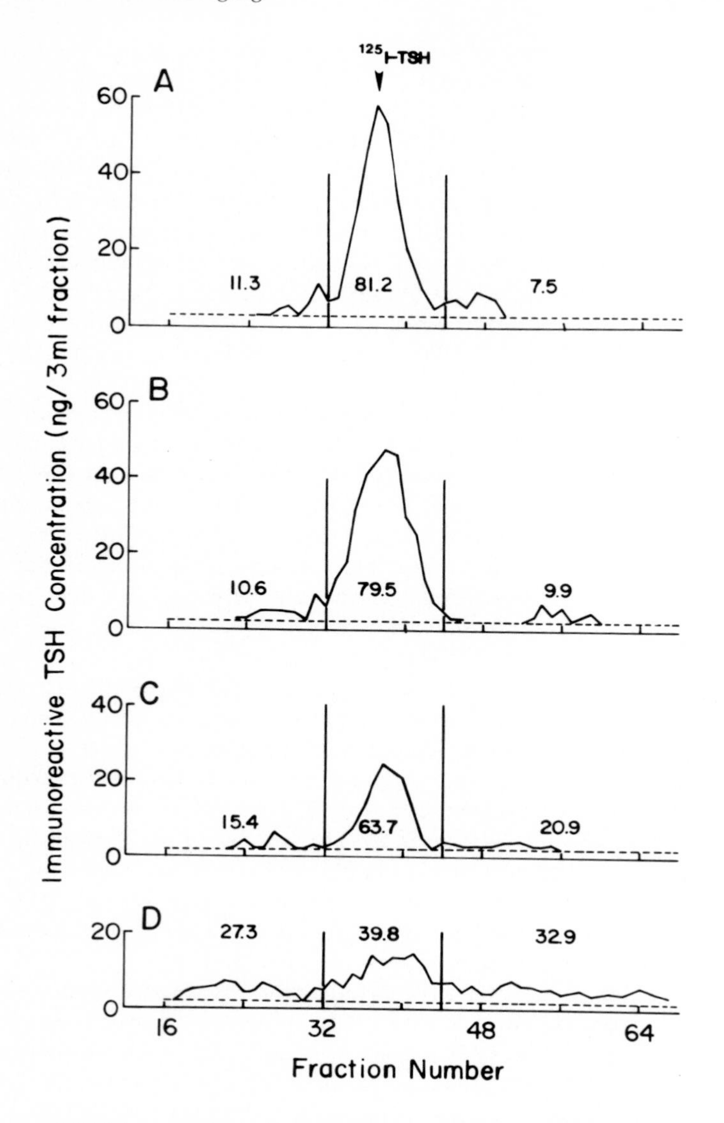

FIGURE 5. Sephadex® G-200 fractionation of glycoprotein serum extracts from rats aged 3 months (A), 12 months (B), 22 months (C), and 30 months (D). The serum extract was the G fraction from the concanavalin A-column. The arrowhead marks the relative elution position of ^{125}I-TSH. The numbers over the graphs are the percentages of IR-TSH within the cut-off lines relative to the total IR-TSH recovered from the column.

strong. Serum IR-TSH was heterogeneous with respect to both molecular size and carbohydrate composition. These forms of IR-TSH require further characterization of their biochemical nature and their physiological significance.

Several conclusions may be drawn from these and other data. However, several facts concerning the thyrotropic activity of the aged hypophysis should be considered first. As evidenced from TSH levels in the blood and pituitary of aged rats,[20,109] and further evidenced by the loss of circadian periodicity of serum TSH,[110] and altered responsiveness of TSH to TRH,[111] thyrotropic activity is probably altered with age. It is also thought that pituitary and circulating TSH in rats are not identical,[112] and that high-molecular weight thyrotropin has impaired biologic activity.[17,18,52] Therefore, it is

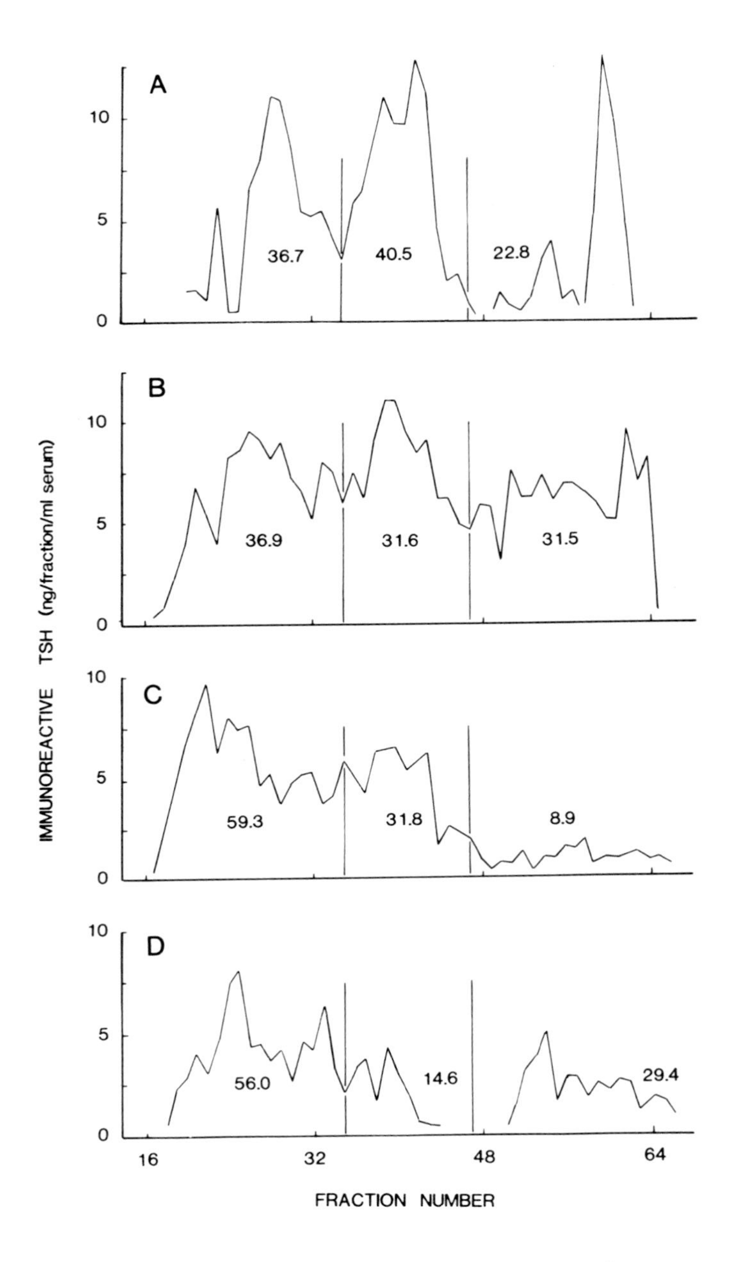

FIGURE 6. Sephadex® G-200 fractionation of nonglycoprotein serum extracts from rats aged 3 months (A), 12 months (B), 22 months (C), and 30 months (D). The serum extract was the NG fractionation from the concanavalin A-column. The numbers over the graphs are the percentages of IR-TSH within the cut-off lines relative to the total IR-TSH recovered from the column.

tempting to speculate that in the aged rat, because of changed thyrotropic function, altered forms of TSH are released into the circulation; these altered forms may have impaired biological identity because of insufficient glycosylation, and may represent incorrectly processed TSH-precursors, or TSH-subunit-precursors.

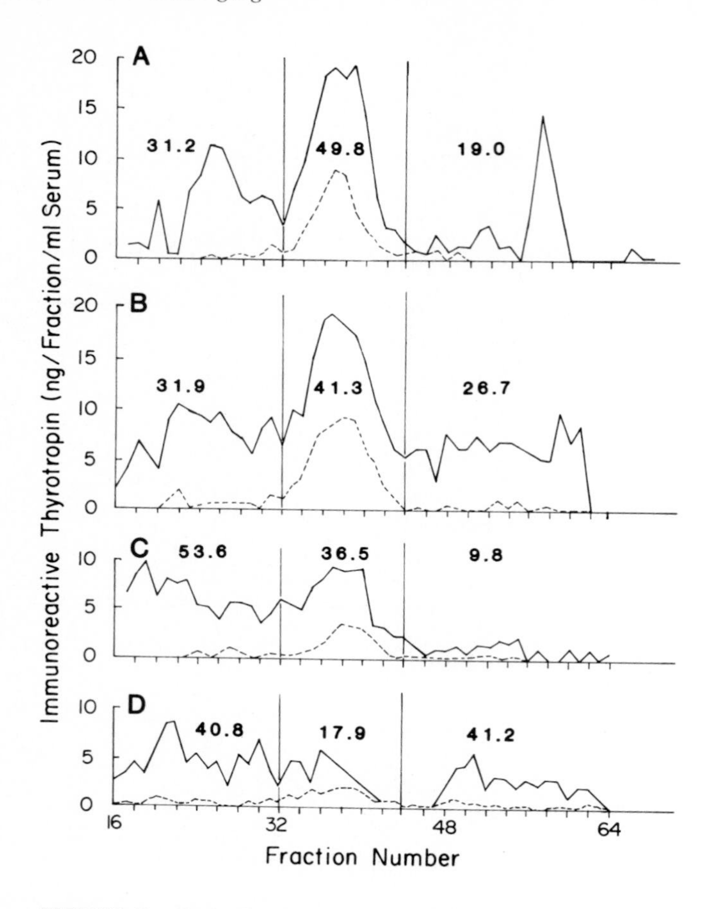

FIGURE 7. Derived Sephadex® G-200 elution profile of whole serum from rats aged 3 months (A), 12 months (B), 22 months (C), and 30 months (D). These curves were calculated from the data already expressed in Figure 5 and Figure 6. The contribution from the G fraction (Figure 5) is shown as a dotted line. The numbers over the graphs are the percentages of IR-TSH within the cut-off lines relative to the total IR-TSH.

V. FINAL REMARKS

In recent years investigators have recognized that not only quantitative parameters but also qualitative parameters must be included in order to understand the complex actions and control mechanisms of hormones. The result has been a greater appreciation of hormone biosynthesis and secretion, and a proliferation of studies into hormone polymorphism. Numerous reports are available to show that polypeptide hormones occur in multiple forms in the circulation and tissue storage depots of normal and diseased individuals. In particular, the pituitary protein hormones are known to possess several types of heterogeneity, some of which have a major influence on immunological and biological identity. For example, glycoprotein hormones appear to be sensitive to carbohydrate composition for their biological potency, but immunological activity is a far more robust property.

Study of altered hormones in the aged organism has not advanced beyond the phenomenological stage. The sparse data base in this field may be indicative of either an absence of a general trend towards increased hormone heterogeneity with senescence or a lack of suitable techniques to make the observations. A further difficulty is the diverse nature of hormone polymorphism that may occur; the data available indicate a different kind of polymorphism for each of the hormones so far investigated. With advancing age, LH shifts to a molecular species that is larger in molecular weight because of increased glycosylation, insulin may accumulate as its unprocessed precursor, proinsulin, and TSH accumulates as high-molecular weight forms that may also be nonglycosylated. In contrast, it has recently been reported that aging causes a decline in the production of the ACTH precursor, pro-opiocortin, and that its decline is not accompanied by major changes in the processing of pro-opiocortin.[26,113] Until more data concerning the occurrence of hormone polymorphism in aged subjects are available, the significance of the phenomenon to the aging process cannot be ascertained.

ACKNOWLEDGMENTS

Supported by research grants from the National Institutes of Health (HD 07430 and AG 00043, P.S.T.) and the Medical Research Council of New Zealand (V.J.C.). All experimental work was done in the laboratory of Professor Paola S. Timiras, Department of Physiology-Anatomy, University of California, Berkeley. Special thanks to P.S.T. for support and encouragement, D. B. Hudson and P. E. Segal for helpful discussion and suggestions, W. R. Klemme for technical assistance. Dedicated to Daisy and Norman Choy.

REFERENCES

1. **Orgel, L. E.**, The maintenance of the accuracy of protein synthesis and its relevance to aging, *Proc. Natl. Acad. Sci. USA*, 49, 517, 1963.
2. **Orgel, L. E.**, Aging of clones of mammalian cells, *Nature (London)*, 243, 441, 1973.
3. **Foote, R. S. and Stulberg, M. P.**, Efficiency and fidelity of cell-free protein synthesis by transfer RNA from aged mice, *Mech. Ageing Dev.*, 13, 93, 1980.
4. **Gershon, D., Gershon, H., Jacobus, S., Reiss, U., and Reznick, A.**, The accumulation of faulty enzyme molecules in aging cells, in *Metabolic Interconversion of Enzymes*, Saltiel, S., Ed., Springer-Verlag, Berlin, 1976, 227.
5. **Reiss, U. and Rothstein, M.**, Heat-labile isozymes of isocitrate lyase from aging *Turbatrix aceti*, *Biochem. Biophys. Res. Commun.*, 61, 1012, 1974.
6. **Reiss, U. and Rothstein, M.**, Age-related changes in isocitrate lyase from the free-living nematode, *Turbatrix aceti*, *J. Biol. Chem.*, 250, 826, 1975.
7. **Reznick, A. Z. and Gershon, D.**, The effect of age on the protein degradation system in the nematode *Turbatrix aceti*, *Mech. Ageing Dev.*, 11, 403, 1979.
8. **Gershon, H. and Gershon, D.**, Detection of inactive enzyme molecules in aging organisms, *Nature (London)*, 227, 1214, 1970.
9. **Gershon, H. and Gershon, D.**, Inactive enzyme molecules in aging mice: liver aldolase, *Proc. Natl. Acad. Sci. USA*, 70, 909, 1973.
10. **Gershon, H. and Gershon, D.**, Altered enzyme molecules in senescent organisms: mouse muscle aldolase, *Mech. Ageing Dev.*, 2, 33, 1973.
11. **Reiss, U. and Gershon, D.**, Rat liver superoxide dismutase purification and age related modifications, *Eur. J. Biochemistry*, 63, 617, 1976.
12. **Mennecier, F. and Drefus, J. C.**, Molecular aging of fructose-biphosphate aldolase in tissues of rabbit and man, *Biochim. Biophys. Acta*, 364, 320, 1974.
13. **Kahn, A., Boivin, P., Viber, M., Cottreau, D., and Dreyfus, J. C.**, Post-translational modifications of human glucose-G-phosphate dehydrogenase, *Biochimie*, 56, 1395, 1974.
14. **Demchenko, A. P. and Orlovska, N. N.**, Age-dependent changes of protein structure. II. Conformational differences of aldolase of young and old rabbits, *Exp. Gerontol.*, 15, 619, 1980.

15. **Sinha, Y. N.**, Molecular size variants of prolactin and growth hormone in mouse serum: strain differences and alterations of concentration by physiological and pharmacological stimuli, *Endocrinology*, 107, 1959, 1980.
16. **Duckworth, W. C. and Kitabchi, A. E.**, The effect of age on plasma proinsulin-like material after oral glucose, *J. Lab. Clin. Med.*, 88, 359, 1976.
17. **Klug, T. L. and Adelman, R. C.**, Evidence for a large thyrotropin and its accumulation during aging in rats, *Biochem. Biophys. Res. Commun.*, 77, 1431, 1977.
18. **Klug, T. L. and Adelman, R. C.**, Age-dependent accumulation of an immunoreactive species of thyrotropin (TSH) which inhibits production of thyroid hormones, *Adv. Exp. Med. Biol.*, 97, 259, 1978.
19. **Conn, P. M., Cooper, R., McNamara, C., Rogers, D. C., and Shoenhardt, L.**, Qualitative change in gonadotropin during normal aging in the male rat, *Endocrinology*, 106, 1549, 1980.
20. **Choy, V. J., Klemme, W. R., Timiras, P. A.**, Variant forms of immunoreactive thyrotropin in aged rats, *Mech. Ageing Dev.*, 19, 273, 1982.
21. **Lang, R. E., Fehm, H. L., Voight, K. H. and Pfeiffer, E. R.**, Two ACTH species in rat pituitary gland, *FEBS Lett.*, 37, 197, 1973.
22. **Yalow, R. S. and Berson, S. A.**, Size heterogeneity of immunoreactive human ACTH in plasma and in extracts of pituitary glands and ACTH-producing thymoma, *Biochem. Biophys. Res. Commun.*, 44, 439, 1971.
23. **Haralson, M. A., Fairfield, S. J., Nicholson, W. E., Harrison, R. W., and Orth, D. N.**, Cell-free synthesis of mouse corticotropin, *J. Biol. Chem.*, 254, 2172, 1979.
24. **Paquette, T. L., Herbert, E., and Hinman, M.**, Molecular weight form of adrenocorticotropic hormone secreted by primary cultures of mouse anterior pituitary, *Endocrinology*, 104, 1211, 1979.
25. **Mains, R. E. and Eipper, B. A.**, Biosynthesis of adrenocroticotrophic hormone in mouse pituitary tumor cells, *J. Biol. Chem.*, 251, 4115, 1976.
26. **Barnea, A., Cho, G., and Porter, J. C.**, Molecular-weight profiles of immunoreactive corticotropin in the hypothalamus of the aging rat, *Brain Res.* 232, 355, 1982.
27. **Reichert, L. E. and Ramsey, R. B.**, Evidence for the existence of a large molecular weight protein in human pituitary tissue having follicle stimulating hormone activity, *J. Clin. Endocrinol. Metab.*, 44, 545, 1977.
28. **Peckham, W. D. and Knobil, E.**, Qualitative changes in the pituitary gonadotropin of the male rhesus monkey following castration, *Endocrinology*, 98, 1061, 1976.
29. **Bogdanove, E. M., Campbell, G. T., Blair, E. D., Mula, M. E., Miller, A. E., and Grossman, G. H.**, Gonad-pituitary feedback involves qualitative change: androgens alter the type of FSH secreted by the rat pituitary, *Endocrinology*, 95, 219, 1974.
30. **Chappell, S. C.**, The presence of two species of FSH within hamster anterior pituitary glands as disclosed by concanavalin A chromatography, *Endocrinology*, 109, 935, 1981.
31. **Einarsson, R. and Skoog, B.**, Use of electrophoresis and gel filtration to characterize human pituitary somatotropin preparations after storage, *Hormone Res.*, 10, 104, 1979.
32. **Stachura, M. E. and Frohman, L. A.**, Growth hormone: independent release of big and small forms from rat pituitary *in vitro*, *Science*, 187, 447, 1974.
33. **Guyda, H. J.**, Heterogeneity of human growth hormone and prolactin secreted in vitro: immunoassay and radioreceptor assay correlations, *J. Clin. Endocrinol. Metab.*, 41, 953, 1975.
34. **Chrambach, A., Yadley, R. A., Ben-David, M., and Rodbard, D.**, Isohormones of human growth hormone. I. Characterization by electrophoresis and isoelectric focusing in polyacrylamide gel, *Endocrinology*, 93, 848, 1973.
35. **Lewis, U. J., Singh, R. N. P., Bonewald, L. F., Lewis, L. J. and Vanderlaan, W. P.**, Human growth hormone: additional members of the complex, *Endocrinology*, 104, 1256, 1979.
36. **Benveniste, R. and Frohman, L. A.** An immunological approach to the study of protein conformational heterogeneity: its application to growth hormone, *Endocrinology*, 102, 198, 1978.
37. **Prentice, L. G. and Ryan, R. J.**, LH and its subunits in human pituitary, serum and urine, *J. Clin Endocrinol. Metab.*, 40, 303, 1975.
38. **Peckham, W. D. and Knobil, E.**, The effects of ovariectomy, estrogen replacement, and neuraminidase treatment on the properties of the adenohypophysial glycoprotein hormones of the rhesus monkey, *Endocrinology*, 98, 1054, 1976.
39. **Liu, T. C., Ax, R. L., and Jackson, G. L.**, Characterization of luteinizing hormone synthesized and released by rat pituitaries *in vitro:* dissociation of immunological and biological activities, *Endocrinology*, 105, 10, 1979.
40. **Robertson, D. M., Puri, V., Lindberg, M., and Diczfalusy, E.**, Biologically active luteinizing hormone (LH) in plasma. V. A re-analysis of the differences in the ratio of biological to immunological LH activities during the menstrual cycle, *Acta Endocrinol.*, 92, 615, 1979.

41. Sakai, C. N. and Channing, C. P., Evidence for alterations in luteinizing hormone secreted in rhesus monkeys with normal and inadequate luteal phases using radioreceptor and radioimmunoassay, *Endocrinology*, 104, 1217, 1979.
42. Wakabayashi, K., Heterogeneity of rat luteinizing hormone revealed by radioimmunoassay and electrofocusing studies, *Endocrinol. Japon.*, 24, 473, 1977.
43. Landefeld, T. D. and Kepa, J., The cell free synthesis of bovine lutropin beta subunit, *Biochem. Biophys. Res. Commun.*, 90, 111, 1979.
44. Gala, R. R., and Hart, I. C., Serum prolactin heterogeneity in the cow and goat, *Life Sci.*, 27, 723, 1980.
45. Suh, H. K. and Frantz, A. G., Size heterogeneity of human prolactin in plasma and pituitary extracts, *J. Clin. Endocrinol. Metab.*, 39, 928, 1974.
46. Sinha, Y. N. and Baxter, S. R., Metabolism of prolactin in mice with a high incidence of mammary tumours: evidence for greater conversion into a non-immunoassayable form, *J. Endocrinol.*, 81, 299, 1975.
47. Lawson, D. M. and Stevens, R. W., Size heterogeneity of pituitary and plasma prolactin: effect of chronic estrogen treatment, *Life Sci.*, 27, 1489, 1980.
48. Lawson, D. M., Size heterogeneity of rat prolactin secreted *in vitro:* effect of incubation time and dopamine, *Proc. Soc. Exp. Biol. Med.*, 165, 364, 1980.
49. Wallis, M., Daniels, M., and Ellis, S. A., Size heterogeneity of rat pituitary prolactin, *Biochem. J.*, 189, 605, 1980.
50. Leung, F. C., Russell, S. M., and Nicoll, C. S., Relationship between bioassay and radioimmunoassay estimates of prolactin in rat serum, *Endocrinology*, 103, 1619, 1978.
51. Erhardt, F. W. and Scriba, P. C., High molecular thyrotropin ("big"-TSH) from human pituitaries: preparation and partial characterization, *Acta Endocrinol.*, 85, 698, 1977.
52. Spitz, I. M., Le Roith, D., Hirsch, H., Carayon, P., Pekonen, F., Liel, Y., Sobel, R., Chorer, Z., and Weintraub, B. Increased high-molecular-weight thyrotropin with impaired biologic activity in a euthyroid man, *New Engl. J. Med.*, 304, 278, 1981.
53. Ponsin, G., Poncet, C., and Mornex, R., Accumulation of a large component related to thyrotropin subunits in the pituitary of thyroidectomized rats, *Biochem. Biophys. Res. Commun.*, 89, 1135, 1979.
54. Pekonen, F., Carayon, P., Amr, S., and Weintraub, B. D., Heterogeneous forms of thyroid-stimulating hormone in mouse thyrotropic tumor and serum: differences in receptor-binding and adenylate cyclase-stimulating activity, *Horm. Metabl. Res.*, 13, 617, 1981.
55. Peterson, V. B., McGregor, A. M., Belchetz, P. E., Elkeles, R. S., and Hall, R., The secretion of thyrotropin with impaired biological activity in patients with hypothalamic-pituitary disease, *Clin. Endocrinol.*, 8, 397, 1978.
56. Yora, T., Matsuzaki, S. Kondo, Y., and Ui, N., Changes in the contents of multiple components of rat pituitary thyrotropin in altered thyroid states, *Endocrinology*, 104, 1682, 1979.
57. Jacobson, G., Roos, P., and Wide, L., Human pituitary thyrotropin: characterization of five glycoproteins with thyrotropin activity, *Biochim. Biophys. Acta*, 490, 403, 1977.
58. Tamura-Takahashi, H. and Ui, N., Purification and properties of whale thyroid-stimulating hormone. III. Properties of isolated multiple components, *Endocrinol. Japon*, 23, 511, 1976.
59. Fawcett, J. S., Dedman, M. L., and Morris, C. J. O. R., The isolation of bovine thyrotropins by isoelectric focussing, *FEBS Lett.*, 3, 250, 1969.
60. Kourides, I. A., Hoffman, B. J., and Landon, M. B., Difference in glycosylation between secreted and pituitary free alpha-subunit of the glycoprotein hormones, *J. Clin. Endocrinol. Metab.*, 51, 1372, 1980.
61. Jacobson, G., Roos, P., and Wide, L., Human pituitary thyrotropin: isolation and recombination of subunit isoforms, *Biochim. Biophys. Acta*, 625, 146, 1980.
62. Golstein-Golaire, J. and Vanhaelst, L., Gel filtration profile of circulating immunoreactive thyrotropin and subunits of myxedematous sera, *J. Clin. Endocrinol. Metab.*, 41, 575, 1975.
63. Kourides, I. A., Weintraub, B. D. and Maloof, F., Large molecular weight TSH-beta: the sole immunoactive form of TSH-beta in certain human sera, *J. Clin. Endocrinol. Metab.*, 47, 24, 1978.
64. Giudice, L. G. and Pierce, J. G., Separation of functional and nonfunctional β-subunits of thyrotropin preparations by polyacrylamide gel electrophoresis, *Endocrinology*, 101, 776, 1977.
65. Lewis, U. J., Singh, R. N. P., Tutwiler, G. F., Sigel, M. B., Vanderlaan, E. F., and Vanderlaan, W. P., Human growth hormone: a complex of proteins, *Rec. Prog. Horm. Res.*, 36, 447, 1980.
66. Pierce, J. G., Eli Lilly lecture: the subunits of pituitary thyrotropin — their relationship to other glycoprotein hormones, *Endocrinology*, 89, 1331, 1971.
67. Wolff, J., Winand, R. J., and Kohn, L. D., The contribution of subunits of thyroid stimulating hormone to the binding and biological activity of thyrotropin, *Proc. Natl. Acad. Sci. USA*, 71, 3460, 1974.

68. **Vaitukaitis, J. L.**, Glycoprotein hormones and their subunits — immunological and biological characterization, in *Structure and Function of the Gonadotropins,* McKern, K. W., Ed., Plenum Press, New York, 1978, 339.
69. **Ross, G. T.**, Clinical relevance of research on the structure of human chorionic gonadotropin, *Am. J. Obstet. Gynecol.,* 129, 795, 1977.
70. **Kourides, I. A., Re, R. N., Weintraub, B. D., Ridgway, E. C., and Maloof, F.**, Metabolic clearance and secretion rates of subunits of human thyrotropin, *J. Clin. Endocrinol.,* 59, 508, 1977.
71. **Hussa, R. O.**, Biosynthesis of human chorionic gonadotropin, *Endocrine Rev.,* 1, 268, 1980.
72. **Weintraub, B. D., Stannard, B. S., Linnekin, D. and Marshall, M.**, Relationship of glycosylation to *de novo* thyroid-stimulating hormone biosynthesis and secretion by mouse pituitary tumor cells, *J. Biol. Chem.,* 255, 5715, 1980.
73. **Vamvakopoulos, N. C., Monahan, J. J., and Kourides, I. A.**, Synthesis, cloning, and identification of DNA sequences complementary to mRNAs for α and β subunits of thyrotropin, *Proc. Natl. Acad. Sci. USA,* 77, 3149, 1980.
74. **Chin, W. W. and Habener, J. F.**, Thyroid-stimulating hormone subunits: evidence from endoglycosidase-H cleavage for late presecretory glycosylation, *Endocrinology,* 108, 1628, 1981.
75. **Bogdanove, E. M., Nolin, J. M., and Campbell, G. T.**, Qualitative and quantitative gonad-pituitary feedback, *Rec. Prog. Horm. Res.,* 31, 567, 1975.
76. **Loh, Y. P. and Gainer, H.**, The role of glycosylation on the biosynthesis, degradation, and secretion of the ACTH-β-lipotropin common precursor and its peptide products, *FEBS Lett.,* 96, 269, 1978.
77. **Channing, C. P., Sakai, C. N., and Bahl, O. P.**, Role of the carbohydrate residues of human chorionic gonadotropin in binding and stimulation of adenosine 3′,5′-monophosphate accumulation by porcine granulosa cells, *Endocrinology,* 103, 341, 1978.
78. **Van Hall, E. V., Vaitukaitis, J. L., Ross, G. T., Hickman, J. W., Ashwell, G.**, Immunological and biological activity of hCG following progressive desialation, *Endocrinology,* 88, 456, 1971.
79. **Krieger, D. T.**, Glandular end organ deficiency associated with secretion of biologically inactive pituitary peptides, *J. Clin. Endocrinol. Metab.,* 38, 964, 1974.
80. **Timiras, P. S., Choy, V. J., Hudson, D. B.**, Neuroendocrine pacemaker for growth, development and ageing, *Age and Ageing,* 11, 73, 1982.
81. **Everitt, A. V.**, Conclusion: aging and its hypothalamic-pituitary control, in *Hypothalamus, Pituitary and Aging,* Everitt, A. V. and Burgess, J. A., Eds., Charles C Thomas, Springfield, Ill., 1976, chap. 34.
82. **Meites, J.**, Changes in neuroendocrine control of anterior pituitary function during aging, *Neuroendocrinol.,* 34, 151, 1982.
83. **Lazarus, L. and Eastman, C. J.**, Assessment of hypothalamic pituitary function in old age, in *Hypothalamus, Pituitary and Aging,* Everitt, A. V. and Burgess, J. A., Eds., Charles C Thomas, Springfield, Ill., 1976, chap. 6.
84. **Everitt, A. V., Seedsman, N. J., and Jones, F.**, The effects of hypophysectomy and continuous food restriction, begun at ages 70 and 400 days, on collagen aging, proteinuria, incidence of pathology and longevity in the male rat, *Mech. Ageing Dev.,* 12, 161, 1980.
85. **Johnson, J. E. and Cutler, R. G.**, Effects of hypophysectomy on age-related changes in the rat kidney glomerulus: observations by scanning and transmission electon microscopy, *Mech. Ageing Dev.,* 13, 63, 1980.
86. **Denckla, W. D.**, Role of pituitary and thyroid glands in the decline of minimal O_2 consumption with age, *J. Clin. Invest.,* 53, 572, 1974.
87. **McKenzie, J. M.**, The bioassay of thyrotropin in serum, *Endocrinology,* 63, 372, 1958.
88. **Mako, M., Block, M., Starr, J., Nielsen, E., Friedman, E., and Rubenstein, A.**, Proinsulin in chronic renal and hepatic failure: a reflection of the relative contribution of the liver and kidney to its metabolism, *Clin. Res.,* 21, 631, 1973.
89. **Cohen, B. J., Anver, M. R., Ringler, D. H., and Adelman, R. C.**, Age-associated pathological changes in male rats, *Fed. Proc. Fed. Am. Soc. Exp. Biol.,* 37, 2848, 1978.
90. **Jacobs, B. B. and Huseby, R. A.**, Neoplasms occurring in aged Fischer rats, with special reference to testicular, uterine and thyroid tumors, *J. Nat. Cancer Inst.,* 39, 303, 1967.
91. **Sass, B., Rabstein, L. S., Madison, R., Nims, R. M., Peters, R. L., and Kelloff, G. J.**, Incidence of spontaneous neoplasms in F344 rats throughout the natural lifespan, *J. Nat. Cancer Inst.,* 54, 1449, 1975.
92. **Coleman, G. L., Barthold, S. W., Osbaldiston, G. W., Foster, S. J., and Jonas, A. M.**, Pathological changes during aging in barrier-reared Fischer 344 male rats, *J. Gerontology,* 32, 258, 1977.
93. **Good, B. F. and Stenhouse, N. S.**, An improved biosasay for TSH by modication of the method of McKenzie, *Endocrinology,* 78, 429, 1966.
94. **Shishiba, Y. and Soloman, D. H.**, A modification of the McKenzie bioassay for long-acting thyroid stimulator (LATS), *J. Clin. Endocrinol. Metab.,* 29, 405, 1969.

95. **Manley, S. W., Bourke, J. R., and Hawker, R. W.,** An in vitro bioassay for thyrotropin (TSH), *Endocrinology,* 84, 1286, 1969.
96. **Bitensky, L., Alaghband-Zadeh, J., and Chayen, J.,** Studies on thyroid stimulating hormone and the long-acting thyroid stimulating hormone, *Clin. Endocrinol.,* 3, 363, 1974.
97. **Berson, S. A. and Yallow, R. S.,** General methodology, in *Methods in Radioimmunoassay of Peptide Hormones,* Yalow, R. S., Ed., North-Holland, Amsterdam, 1976, chap. 1.
98. **Vanhaelst, L., Bonnyns, M., and Golstein-Golaire, J.,** Pituitary TSH in normal subjects and in patients with asymptomatic atrophic thyroiditis: evidence for its immunological identity, *J. Clin. Endocrinol. Metab.,* 41, 115, 1975.
99. **Pekonen, F. and Weintraub, B. D.,** Thyrotropin receptors on bovine thyroid membranes: two types with different affinities and specificties, *Endocrinology,* 105, 352, 1979.
100. **Pekonen, F. and Weintraub, B. D.,** Interaction of crude and pure chorionic gonadotropin with the thyrotropin receptor, *J. Clin. Endocrinol. Metab.,* 50, 280, 1980.
101. **Odell, W. D. and Utiger, R. D.,** Thyroid-stimulating hormone (TSH), in *Methods in Radioimmunoassay of Peptide Hormones,* Yalow, R. S., Ed., North-Holland, Amsterdam, 1976, chap. 9.
102. **Kristiansen, T.,** Group-specific separation of glycoproteins, in *Methods in Enzymology,* Vol. 34, Jacoby, W. B. and Wilcheck, M., Eds., Academic Press, New York, 331, 1974.
103. **Asberg, K. and Porath, J.,** Group-specific adsorption of glycoproteins, *Acta Chem. Scand.,* 24, 1839, 1970.
104. **Dufau, M. L., Tsuruhara, T., and Catt, K. J.,** Interaction of glycoprotein hormones with agarose-concanavalin A, *Biochim. Biophys. Acta,* 278, 281, 1972.
105. **Ui, N., Tamura-Takahashi, H., Yora, T., and Condliffe, P. G.,** Bioaffinity chromatography of thyrotropin using immobilized concanavalin A, *Biochim. Biophys. Acta,* 497, 812, 1977.
106. **Bloomfield, G. A., Faith, M. R., and Pierce, J. G.,** Sepharose-linked concanavalin A in the purification and characterization of glycoprotein hormones of the bovine pituitary, *Biochim. Biophys. Acta,* 533, 371, 1978.
107. **Louvet, J.-P., Nisula, B. C., Ross, G. T.,** Method for extraction of glycoprotein hormones from plasma for use in radioimmunoassays, *J. Lab. Clin. Med.,* 86, 883, 1975.
108. **Lowry, O. M., Rosebrough, N. J., Farr, A. L., and Randall, R. J.,** Protein measurement with the Folin phenol reagent, *J. Biol. Chem.,* 193, 265, 1951.
109. **Valueva, G. V. and Verzhikovskaya, N. V.,** Thyrotropic activity of hypophysis during aging, *Exp. Gerontol.,* 12, 97, 1977.
110. **Klug, T. L. and Adelman, R. C.,** Altered hypothalamic-pituitary regulation of thyrotropin in male rats during aging, *Endocrinology,* 104, 1136, 1979.
111. **Chen, H. J. and Walfish, P. G.,** Effects of age and testicular function on the pituitary-thyroid system in male rats, *J. Endocrinol.,* 82, 53, 1979.
112. **Blackman, M. R., Gershengorn, M. C., and Weintraub, B. D.,** Excess production of free alpha subunits by mouse pituitary thyrotropic tumor cells in vitro, *Endocrinology,* 102, 499, 1978.
113. **Barnea, A., Cho, G., and Porter, J. C.,** A reduction in the concentration of immunoreactive corticotropin, melanotropin and lipotropin in the brain of the aging rat, *Brain Res.,* 232, 345, 1982.

Chapter 8

COLLAGEN AND AGING

Donald J. Cannon

TABLE OF CONTENTS

I. Introduction 162

II. Collagen 162
  A. Physical Changes 162
  B. Cross-Linking 163
  C. Enzymatic Digestion 165
  D. Biosynthesis 165

III. Future Studies 166

Acknowledgment 166

References 166

I. INTRODUCTION

Connective tissue has long held the interest of investigators seeking to unravel the biochemical processes of aging. The supportive tissues of the body may not be directly involved in biological aging but may simply reflect cellular changes with age. On the other hand, there are many morphological, physical, and biochemical studies that indicate that changes in connective tissue matrices are age-related, yet these are difficult to attribute solely to cellular changes. The role that connective tissue changes may play in organismic aging is unknown although many hypotheses have been proposed.[1] None of these are entirely satisfactory, however changes with age in the connective tissue protein, collagen are firmly established[2] and undoubtedly further studies based on advances in methodology and in our knowledge of this key structural protein will clarify the role of connective tissue in the aging process.

Certainly connective tissue is subjected to continuous stress and deformation during the life of the average adult and to numerous insults that require repair and adjustment to the connective tissue architecture. These phenomena have to be assessed in determining the role of connective tissue during aging. In addition, changes in connective tissue with age that can alter or impede the physiological function of a particular tissue should be differentiated from pathological processes that arise from recent or long-term result.

Studies of connective tissue encompass the synthesis, structure, relationships, and catabolism of the many macromolecular constituents that are found in inter- and extracellular matrices. Age-related connective tissues studies have concentrated chiefly on the protein collagen and its cross-links. This chapter will cover some methods employed to investigate age-related changes in collagen.

II. COLLAGEN

Our basic knowledge of the biochemistry of collagen can be found in recent texts[3,4] and reviews[5-8] and in addition to the general and aging literature there are specialized journals covering recent connective tissue developments.

A. Physical Changes

Collagen has been the most extensively studied component of connective tissue. Initial observations of the increased stability of collagen with age came from the studies of Verzar.[9] Strips of skin or tendon were immersed in warm physiological saline, one end was clamped and the other attached to a kymograph. Contraction was measured as a function of increased temperature. Verzar developed the concept of isotonic contraction by weighting down the tissue to prevent contraction at temperatures above that needed to develop rapid shrinkage (shrinkage temperature, T_s). Rat tail tendons from old (33 month) animals required over 10 times the weight of young (2 month) animals to prevent contraction. A modification of this technique which measures hydrothermal isometric tension has recently been developed.[10] Strips of dissected, shaved skin are clamped in an isometric tension device (see Figure 1 in Reference 10) and subjected to a linear temperature increase in unbuffered saline to prevent tissue swelling. Using this technique, the rate of tension development, the maximum tension and peak tension temperature were correlated with the maturation of reductive cross-links and the age of donor animals in a wide variety of species.[11] The relation of these latter studies to collagen cross-links and aging will be discussed again later.

Connective tissue, rich in collagen, has been examined as a function of the age of the donor animal by a variety of other physical techniques. These have demonstrated increased crystalline organization, decreased water content, and increased tensile strength with donor age.[12-14] These studies have been supplemented by measurements of the solubility of collagen in buffered salt solutions, acetic acid, and various denaturants.[15-17] In general, tissue is dissected free of extraneous fat, fascia, hair, etc., reduced to small pieces by mechanical or manual chopping and dispersed by homogenization or grinding in the solvent of choice. All procedures are done at 4°C and the inclusion of protease inhibitor(s) and an antibacterial agent is necessary to prevent degradation of collagen. Collagenous tissues swell in acid and frequent homogenization during acid solubilization (2 to 6 days) is recommended. Solubilized material is harvested by centrifugation and further purification is achieved by selective precipitation (unique to collagen) in sodium chloride and dialysis against neutral phosphate buffer.[18] The final product should be dialyzed free of salt, and an aliquot assayed for hydroxyproline[19] and the collagen content calculated.[20] There are numerous procedures for the analysis of hydroxyproline[21] that are designed for different tissues or fluids, different levels of hydroxyproline and the presence of radioactive hydroxyproline and proline. Each must be assessed for a particular experimental design.

The above solubilization procedures have been employed with tissue that consist solely of collagen isotype I, which is the most abundant isotype and is a trimer of two chains $(\alpha_1 I)_2(\alpha_2)$. Bone collagen consists entirely of this isotype but must first be decalcified in buffered EDTA prior to solubilization of collagen.[22] Hyaline cartilage consists solely of collagen isotype II, a trimer of a single chain $(\alpha_1 II)_3$ and is resistant to chemical solubilization.[23] Digestion with enzymes[24,25] is necessary to release cartilage collagen either before or after the chemical removal of proteoglycans which are rich in this tissue. Most other tissues (lung, kidney, skin, aorta, etc.) are composites of two to four collagen isotypes in which I and III predominate.[7] The latter is a single chain trimer $(\alpha_1 III)_3$ exclusively associated with I in interstitial tissues. The remaining isotypes, IV and V, are found in much smaller quantities in select tissues with basement membranes. Solubilization of multitype collagen tissues is achieved with pepsin[26] and the individual types can be fractionated by selective salt precipitation.[27] Since the level of hydroxyproline varies between isotypes the relevant amounts of types are estimated[28] or quantitated by radiolabeling followed by digestion and chromatography.[29]

The net result of physical chemical studies and the solubility of collagen from select tissues (e.g., tendon) indicated an increased stability of collagen with age which was attributed to increased cross-linking of the collagen molecule.

B. Cross-Linking

Present knowledge of collagen cross-linking has been reviewed[30] and will be only briefly described here. Lysyl and hydroxylysl residues, predominately located in the amino and carboxy terminal regions of the collagen chains, are oxidized extracellularly by the enzyme lysyl oxidase to form the aldehydes allysine and hydroxyallysine. These aldehydes condense spontaneously when in apposition with other residues of lysine, hydroxylysine, and histidine or a second residue of allysine or hydroxyallysine to produce intra- and intermolecular bonds that (for intermolecular bonds) can potentially involve the binding of 2,3, or 4 peptide chains. Most of the adducts formed from these chemical condensations have been isolated and identified by reduction with radiolabeled sodium borohydride.[31] In addition to labeling the compounds of interest, reduction stabilizes them (and the collagen molecule) permitting subsequent acid and alkaline hydrolysis, chromatography, and identification. Certain of the cross-links are destroyed or rearranged by strong acid[32] and certain cross-links derived from hydroxylysine are glycosylated thus necessitating alkaline hydrolysis of reduced, unadulterated tissue.[33] Separation of the cross-links is achieved by ion-exchange chromatography on resins suitable for amino acid analysis and employing a variety of elution buffer systems.[33-35] Collected fractions are assayed for radioactivity and by comparison with the elution of known isolated or synthesized standards, and the specific activity of tritiated sodium borohydride, the amount of individual cross-links can be calculated.

The predominant reducible cross-link(s) varies in different tissues[30] due mainly to variable post-translational hydroxylation of lysyl residues. In addition, there are changes in the cross-link distribution with development and age.[33,36] In most tissues there is a reduction with age in total reducible cross-links which would appear to contradict some of the observations discussed earlier. In fact, the reducible cross-links appear to become stabilized with age and the presence of heat stabile cross-links in older tissues has been demonstrated by increases in the temperature of maximally developed tension in old and young tissue.[10] With age then the reducible cross-links of collagen became heat, acid, and alkali stable and nonreducible. This conclusion has led to searches for in vivo reduced cross-links but none have been satisfactorily demonstrated.[37]

The unique peptide fingerprint of collagen, created by the presence of methionine residues which are cleaved by cyanogen bromide, has been studied in old and young bovine tendon to detect increased cross-linking.[38,39] No changes were observed in cross-linked peptides with aging. Conflicting interpretations concerning changes that occur with maturation are unresolved and may be unique to tendon tissue.

Investigations into the presence of nonreducible mature cross-links that confer on collagens the properties described above have centered on reactions involving the structure of known cross-links. In vivo oxidation of the major cross-link of skin collagen (hydroxylysinonorleucine) has been proposed as a pathway for stabilization of this cross-link to a structure that is not reducible and that produces 2-amino adipic acid following acid hydrolysis.[40] The latter compound was demonstrated in hydrolysates of skin, was derived from peptidyl lysine and was increased in old vs. young bovine skin.[40]

Since the quantities of aminoadipic acid isolated are so low, gram quantities of tissue, of known hydroxyproline content, are hydrolyzed in 6 *N* HCl. After evaporation of HCl, amino acids are partially separated (0.75 *M* NaOH) by displacement chromatography on cation exchange columns in the presence of exogenous ^{14}C-aminoadipic acid. Radioactive monitoring of the column effluent allows the total hydrolysate to be reduced to an area of interest which can then be quantitated by high resolution cation exchange chromatography.[40] From the elution position, standard values for aminoadipic (ninhydrin or fluorescence), and correcting for losses encountered in the isolation steps, values in different tissues can be assessed. (Note: aminoadipic acid forms a lactam at acid pH). Further evidence for the presence of the nonreducible adipyl hydroxylysine cross-link will have to come from large-scale isolations of cross-linked peptides containing this unique component.

A fluorescent cross-linking amino acid, pyridinoline, has recently been isolated from collagen.[41] This compound is a 3-hydroxypyridinium derivative containing three amino and three carboxyl groups and thus capable of linking three collagen chains.[42] It is proposed that pyridinoline arises from the condensation of two preexisting reducible cross-links to form a nonreducible fluorescent cross-link.[43] Despite skepticism that pyridonoline was an artifact of isolation[44] evidence indicates that the compound is a native constituent of collagen,[45,46] although precautions must be taken because of the instability of pyridinoline to light.[47] Pyridinoline is found in bone,[42] cartilage,[43] dentin,[48] and tendon[42] but not skin. Its presence in urine supports a role for its assessment in bone diseases.[49] It is present in dense connective tissue at high levels compared to the reducible cross-links and is especially rich in old tissue. Changes in the properties of collagen with age in these tissues have been attributed to the presence of pyridinoline, although further investigation is needed. It is apparent that the pathway to a nonreducible stable, age-related collagen cross-link could be different in different tissues.

The addition of a histidyl residue to the aldol condensation product (an intramolecular cross-link arising from the condensaton of two allysyl residues) results in the formation of dehydro histidino-hydroxymerodesmosine.[50] This cross-link might also be a precursor candidate for a nonreducible, stable cross-link of aged tissues. Although, like pyridinoline, this compound was thought to be an artifact of in vitro reduction,[51] recent evidence indicates that it is an integral part of the soft tissue cross-linking repertoire.[52]

The formation of reducible Schiff bases from the condensation of monosaccharide aldehyde with the ε-amino group of collagen lysyl residues increases with age.[53] This is not a reflection of cross-link formation but is similar to the increased levels of hemoglobin glycosylation observed in the aging red blood cell.[54] A recent report has shown that the levels of ketoamine-linked glucose in insoluble collagen increases with age.[55] These changes were observed in skin and tendon with dramatic changes in the latter presumably due to decreased levels of collagen turnover in tendon vs. skin. As yet there is no correlation between the formation of the radioactive *N*-glycosylamines, which are detected following borohydridge reduction and chromatography, and the quantitation of ketoamine-linked glucose residues. The latter are determined on insoluble collagen following a 4.5-hr boiling in 1:0 *N* oxalic acid.[55] The released ketoamine glycosyl residues are quantitated by conversion to their furfural derivative. Different methodologies may be detecting the same post-translational modification of collagen and although glycosylation of collagen is correlated with age and insolubility, it is a marker of tissue age rather than a clue to cross-linking changes. In fact, increased levels of ketoamine glucosyl residues seen in diabetic connective tissue have been attributed to accelerated collagen aging in these tissues and have been implicated in the connective tissue complications seen in diabetics.[55]

C. Enzymatic Digestion

Another interesting marker of collagen aging is the resistance of insoluble collagen to digestion with bacterial collagenase and trypsin.[56] Rates of enzymatic digestion decrease progressively and correlate well with postmature chronological age.[57] The rate of digestion is species specific and has been observed in tendon, dura mater, fascia, and myocardial tissue.[58] In this procedure, collagen, dissected free of nonconnective tissue is washed in neutral buffer (0.01 *M* Tris, pH 7.4), delipidated in n-hexane, rewetted in neutral buffer and homogenized in liquid nitrogen at −60°C. The homogenate is extracted twice each in 0.15 *M* NaCl and then 1.0 *M* NaCl (buffered) at 4°C. The resulting insoluble collagen is homogenized in water, lyophilized, and extracted again in hexane and relyophilized. Collagen content is determined by hydroxyproline analysis and the insoluble tissue is thoroughly suspended in buffer (pH 7.8) containing 0.1 *M* $CaCl_2$. Enzymatic digestion at 37°C is monitoried by alkaline (0.01 *M* NaOH) titration for 1 hr at an enzyme:substrate ratio of 1:50. Critical steps in the above procedure are careful preparation of insoluble collagen, standardized dispersion of substrate during proteolysis and proper use of $CaCl_2$. Concentration of the latter is 1.0 *M* for optimum results with trypsin digestion of insoluble collagen.[57]

At present, it is assumed that resistance to enzymatic digestion in postmature collagen is related to changes in collagen cross-links. With the development of better techniques to identify and quantitate postmature collagen cross-links this hypothesis may be tested.

D. Biosynthesis

The biosynthesis of the collagen molecule consists of a series of unique intra- and extracellular steps, the details of which have been the subject of recent reviews.[5,6] The effects of aging on the biosynthesis of collagen have not been systematically studied.

We do know that with age the rate of synthesis and degradation of collagen decreases.[5] These observations, derived mostly from enzyme activity measurements of skin, are consistent with the increased stability (low turnover) of collagen that occurs with time. Many in vitro aging studies use fibroblasts as models of aging, yet despite the potential of this cell to synthesize collagen, few studies have investigated the quality and quantity of collagen synthesis with age (population doublings) in these cells and those that have describe conflicting conclusions.[59,60]

III. FUTURE STUDIES

Most studies of age-related changes in collagen have used, as starting material, stable supportive tissues (e.g., tendon) consisting solely of type I collagen. For reasons of practical methodology and uncomplicated data interpretation this was necessary. Given that collagen reflects age-related biological processes and does not initiate them, then future research should concentrate on more complex connective tissues. Whether the approach is to study biosynthetic, post-translational, or other mechanisms, age-related studies of collagen should encompass the interaction of collagen with noncollagenous macromolecules, changes within the family of collagen isotypes, and the integrity of collagen biosynthesis and catabolism, to name a few. These types of studies will involve tissues which demonstrate age-associated pathological changes and discrimination between these processes and aging will be a formidable task.

ACKNOWLEDGMENTS

The author is grateful for the continued support of the Veterans Administration Research Service.

REFERENCES

1. **Strehler, B. L.,** *Time, Cells and Aging,* 2nd ed., Academic Press, New York, 1977.
2. **Hall, D. A.,** *The Aging of Connective Tissue,* Academic Press, New York, 1976.
3. **Ramachandran, G. N. and Reddi, A. H.,** *Biochemistry of Collagen,* Plenum Press, New York, 1976.
4. **Viidik, A. and Vuust, J.,** *Biology of Collagen,* Academic Press, London, 1980.
5. **Prockop, D. J., Kivirikko, K. I., Tuderman, L., and Guzman, N. A.,** The biosynthesis of collagen and its disorders, *N. Engl. J. Med.,* 301 (13), 75, 1979.
6. **Fessler, J. H. and Fessler, L. I.,** Biosynthesis of procollagen, *Ann. Rev. Biochem.,* 47, 129, 1978.
7. **Bornstein, P. and Sage, H.,** Structurally distinct collagen types, *Ann. Rev. Biochem.,* 49, 957, 1980.
8. **Bornstein, P. and Traub, W.,** The chemistry and biology of collagen, in *The Proteins,* 3rd ed., Neurath, H. and Hill, R., Eds., Academic Press, New York, 1979, 411.
9. **Verzar, F.,** Aging of the collagen fiber, *Int. Rev. Conn. Tissue Res.,* 2, 240, 1963.
10. **Allain, J. C., LeLous, M., Bazin, S., Bailey, A. J., and Delaunay, A.,** Isometric tension developed during heating of collagenous tissue, *Biochim. Biophys. Acta,* 533, 147, 1978.
11. **Cohen-Solal, L., LeLous, M., Allain, J. C., and Meunier, F.,** Absence of maturation of collagen crosslinks in fish skin, *FEBS Lett.,* 123, 282, 1981.
12. **Viidik, A.,** Connective tissues — possible implications of the temporal changes for the aging process, *Mech. Ageing Dev.,* 9, 267, 1979.
13. **Viidik, A.,** *Aging of Connective and Skeletal Tissue,* Engle, A. and Larsson, T., Eds., Nordiska Bokhandelns Forlag, Stockholm, 1969, 125.
14. **Vogel, H. G.,** Influence of maturation and aging on mechanical and biochemical properties of connective tissue in rats, *Mech. Ageing Dev.,* 14, 283, 1980.
15. **Maekawa, T., Rathinasamy, T. K., Altman, K. I., and Forbes, W. F.,** Changes in collagen with age, *Exp. Gerontol.,* 5, 177, 1970.

16. Cannon, D. J. and Davison, P. F., Crosslinking and aging in rat tendon collagen, *Exp. Gerontol.*, 8, 51, 1973.
17. Wirtschafter, Z. T. and Bentley, J. P., The extractable collagen of lathyritic rats with relation to age, *Lab. Invest.*, 11, 316, 1962.
18. Gallop, P. M. and Seifter, S., Preparation and properties of soluble collagen, *Method Enzymol.*, 6, 635, 1963.
19. Rojkind, M. and Gonzalez, E., An improved method for determining specific radioactivities of proline-^{14}C and hydroxyproline-^{14}C in collagen and noncollagenous proteins, *Anal. Biochem.*, 57, 1, 1974.
20. Prockop, D. J. and Udenfriend, S., A specific method for the analysis of hydroxyproline in tissues and urine, *Anal. Biochem.*, 1, 228, 1960.
21. Jamall, I. S., Fimelli, V. N., and Que Hee, S. S., A simple method to determine nanogram levels of 4-hydroxyproline in biological tissues, *Anal. Biochem.*, 112, 70, 1981.
22. Miller, E. J. and Martin, G. M., The collagen of bone, *Clin. Orthop.*, 59, 195, 1968.
23. Miller, E. J. and Matukas, V., Chick cartilage collagen. A new type of α_1 chain not present in bone or skin of the species, *Proc. Natl. Acad. Sci. USA*, 64, 1264, 1969.
24. Miller, E. J., Structural studies on cartilage collagen employing limited cleavage and solubilization with pepsin, *Biochemistry*, 11, 4903, 1972.
25. Simunek, Z. and Muir, H., Changes in the protein polysaccharides of pig articular cartilage during prenatal life, development and old age, *Biochem. J.*, 126, 515, 1972.
26. Hong, B. S., Davison, P. F., and Cannon, D. J., Isolation and characterization of a distinct type of collagen from bovine fetal membranes and other tissues, *Biochemistry*, 18, 4278, 1979.
27. Trelstad, R. L., Catanese, V. M., and Ruben, D. F., Collagen fractionation, *Anal. Biochem.*, 71, 114, 1976.
28. Epstein, E. H., Jr., $[\alpha 1(III)]_3$ human skin collagen, *J. Biol. Chem.*, 249, 3225, 1974.
29. Reiser, K. M. and Last, J. A., Quantitation of specific collagen types from lungs of small animals, *Anal. Biochem.*, 104, 87, 1980.
30. Tanzer, M. L., Cross-linking, in *Biochemistry of Collagen*, Ramachandran, G. N. and Reddi, A. H., Eds., Plenum Press, 1976, 137.
31. Cannon, D. J. and Davison, P. F., A stabilized tris (hydroxymethyl) aminomethane adduct in reduced collagen, *Conn. Tissue Res.*, 4, 187, 1976.
32. Davison, P. F., Cannon, D. J., and Anderson, L. P., The effects of acetic acid on collagen crosslinks, *Conn. Tissue Res.*, 1, 205, 1972.
33. Cannon, D. J. and Davison, P. F., Aging and crosslinking in mammalian collagen, *Exp. Aging Res.*, 3, 87, 1977.
34. Bailey, A. J., Peach, C. M., and Fowler, L. J., Chemistry of the collagen crosslinks, *Biochem. J.*, 117, 819, 1970.
35. Mechanic, G. L., A two column system for complete resolution of $NaBH_4$-reduced crosslinks from collagen, *Anal. Biochem.*, 61, 355, 1974.
36. Robins, S. P., Shimokomaki, M., and Bailey, A. J., The chemistry of the collagen crosslinks. Age related changes in the reducible components of intact bovine collagen fibers, *Biochem. J.*, 131, 771, 1973.
37. Bailey, A. J. and Peach, C. M., Chemistry of collagen crosslinks: the absence of reduction of dehydrolysinonorleucine and dehydroxylysinonorleucine, *in vivo*, *Biochem. J.*, 121, 257, 1971.
38. Davison, P. F., Bovine tendons: aging and collagen crosslinking, *J. Biol. Chem.*, 253, 5635, 1978.
39. Light, N. D. and Bailey, A. J., Changes in crosslinking during aging in bovine tendon collagen, *FEBS Lett.*, 97, 183, 1979.
40. Bailey, A. J., Ranta, M. H., Nicholls, A. C., Partridge, S. M., and Elsden, D. F., Isolation of α-aminoadipic acid from mature dermal collagen and elastin, *Biochem. Biophys. Res. Commun.*, 78, 1403, 1977.
41. Fujimoto, D., Akiba, K., and Nakamura, N., Isolation and characterization of a fluorescent material in bovine achilles tendon collagen, *Biochem. Biophys. Res. Commun.*, 76, 1124, 1977.
42. Fujimoto, D., Moriguchi, T., Ishida, T., and Hayashi, H., The structure of pyridinoline, a collagen crosslink, *Biochem. Biophys. Res. Commun.*, 84, 52, 1978.
43. Eyre, D. R. and Oguchi, H., The hydroxypyridinium crosslinks of skeletal collagen: their measurement, properties and a proposed pathway of formation, *Biochem. Biophys. Res. Commun.*, 92, 403, 1980.
44. Elsden, D. F., Light, N. D., and Bailey, A. J., An investigation of pyridinoline, a putative collagen crosslink, *Biochem. J.*, 185, 531, 1980.
45. Fujimoto, D., Evidence for natural existence of pyridinoline crosslink in collagen, *Biochem. Biophys. Res. Commun.*, 93, 948, 1980.
46. Tsuchikura, O., Gotoh, Y., and Saito, S., Pyridinoline fluorescence in cyanogen peptides of collagen, *Biochem. Biophys. Res. Commun.*, 102, 1203, 1981.

47. **Sakura, S. and Fujimoto, D.,** Electrochemical behavior of pyridinoline, a crosslinking amino acid of collagen, *J. Biochem.,* 89, 1541, 1981.
48. **Kuboki, Y., Tsuzaki, M., Sasaki, S., Liu, C. F., and Mechanic, G. L.,** Location of the intermolecular crosslinks in bovine dentin collagen, solubilization with trypsin and isolation of cross-link peptides containing dihydroxylysinonorleucine and pyridinoline, *Biochem. Biophys. Res. Commun.,* 102, 119, 1981.
49. **Gunja-Smith, Z. and Boucek, R. J.,** Collagen cross-linking compounds in human urine, *Biochem. J.,* 197, 759, 1981.
50. **Tanzer, M. L., Housley, T., Berube, L., Fairweather, R., Franzblau, C., and Gallop, P. M.,** Structure of two histidine-containing cross-links from collagen, *J. Biol. Chem.,* 248, 393, 1973.
51. **Robins, S. P. and Bailey, A. J.,** The chemistry of the collagen cross-links. The characterization of fraction C, a possible artifact produced during the production of collagen fibers with borohydride, *Biochem. J.,* 135, 657, 1973.
52. **Bernstein, P. H. and Mechanic, G. L.,** A natural histidine-based minimum cross-link in collagen and its location, *J. Biol. Chem.,* 255, 10414, 1980.
53. **Robins, S. P. and Bailey, A. J.,** Relative stabilizes of the intermediate reducible crosslinks present in collagen fibers, *FEBS Lett.,* 33, 167, 1973.
54. **Fitzgibbons, J. F., Koler, R. D., and Jones, R. T.,** Red cell age-related changes of hemoglobins A_{1a+b} and A_{1c} in normal and diabetic subjects, *J. Clin. Invest.,* 58, 820, 1976.
55. **Schnider, S. L. and Kohn, R. R.,** Glucosylation of human collagen in aging and diabetes mellitus, *J. Clin. Invest.,* 66, 1179, 1980.
56. **Hamlin, C. R. and Kohn, R. R.,** Evidence for progressive age-related structural changes in post-mature human collagen, *Biochim. Biophys. Acta,* 236, 458, 1971.
57. **Hamlin, C. R., Luschin, J. H., and Kohn, R. R.,** *Exp. Gerontol.,* 13, 415, 1978.
58. **Hamlin, C. R., Luschin, J. H. and Kohn, R. R.,** Aging of collagen: comparative rates in four mammalian species, *Exp. Gerontol.,* 15, 393, 1980.
59. **Paz, M. A., and Gallop, P. M.,** Collagen synthesized and modified by aging fibroblasts in culture, *In Vitro,* 11, 302, 1975.
60. **Kontermann, K. and Bayreuther, K.,** The cellular aging of rat fibroblasts in vitro is a differentiation process, *Gerontology,* 25, 261, 1979.

INDEX

A

N-Acetylglucosamine, 139
N-Acetylneuraminic acid, 139
Acid phosphatase, 122
Acid-urea/SDS polyacrylamide gel electrophoresis, 101, 102, 104
Acid-urea/triton-acid urea, 104
Acrylamide, 96
Acrylamide ampholine gel, 115
ACTH, see Adrenocorticotropic hormone
(1)2′,5′ Adenosine diphosphate, 124
Adrenocorticotropic hormone (ACTH), 78, 155
Affinity chromatography, 93, 95, 146
Agarose, 93, 96
Aging process
  cellular changes, 82
  hormone alterations, 136, 140, 154
  supportive tissue changes, 162
  TPI abnormalities, 27, 28
Alanine, 13, 25, 121
Aldolase(s)
  altered, 6, 7, 12
  B in liver, 114, 127, 128
  post-translational modifications in eye lens, 114, 125—127
  purification, 125, 126
Alkaline phosphatase, 122
Alpha amino isobutyric acid, 45
Amido black, 102, 105
Amino acid analogues
  estimating fidelity and, 36
  feeding to induce error, 51
  radiolabeled, purity of, 45, 46
Amino acids, see also specific amino acids
  C and N terminal groups, 6, 7
  composition in peptides, 17
  high sensitivity analysis, 10, 11, 21, 22
  misincorporation, 36
  primary fragmentation, 12, 13
  protein degradation and, 56, 70, 72, 73, 74
  starvation, 52
  substitution, 48
Amino acylation, 50
Amino adipic acid, 164
α-Aminobutyric acid, 25
Amino ethyl cysteine, 25
Aminopeptidase, 22
Aminopolystyrene resin, 25
Amino terminal determination, 22, 23
Aminotransferase, 74
Ammonium hydroxide, 16
Ammonium sulfate precipitation, 3
Ampholytes, 96, 99
Analytical separation techniques, see also specific techniques, 114
Animals, choice of, see also specific animals, 141, 142
Anterior pituitary hormones, 136, 137, 140, 141
Antibody-enzyme ratios, 47
Antibody precipitation, 47, 48
Antigenic cross-reactivity, 136
Arginine, 43, 58
Arginine free medium, 77
Arginine terminal peptides, 23, 25
Argininyl bonds, 12
Asparagine (ASP)
  bond cleavage, 12, 13, 28
  quantitation, 23, 24
Autoradiography, 105, 106
Azetidine carboxylic acid, 52

B

Back hydrolysis in microsequencing, 25, 26
Bacterial systems as models, see also specific bacteria, 36
Bacteriophage DNA replication, 50
Base substitution mutations, 36
Benzene-ethylacetate, 25
Beta subunits, glycoprotein, 139, 144, 145, 150, 151
Bioassays, see also specific bioassays, 143, 144
Blood cells, see specific cells
Bromelain, 122
Buoyant density centrifugation, 85
Butadione, 13

C

Canavanine, 52, 77
Carbohydrate composition, see Glycosylation
Carboiimides, 25
Carbonate labeling, 57, 58, 74
Carboxyl terminal determination, 23, 121
Carboxypeptidase(s), 23, 121, 122
Catalase, 63
Cell cycle traverse, 95
Cell fractionation, 84
Cellulose thin-layer plates, 11
Centrifugation
  nuclear chromatin isolation and, 84, 85, 99
  sucrose density gradient, 93
Cervical carcinoma culture cells, 83
Cesium salts, 85
Characteristic time, 61
Charge substitutions, 48
Chase medium, 72—75
1-Chlorobutane, 25
Chromatin
  isolation, 82—85
  NaCl-urea treatment, 89, 90
  structure, 95
Chromatography, see also specific type of chromatography

affinity, 6, 10
amino terminal determination and, 23
high-pressure liquid, 16
homology peptide mapping and, 21
purification of altered enzymes, 5
silica gel, 26
thin-layer cellulose, 17
Chromosomal proteins, nonhistone, see also Histones; Proteins
analysis by polyacrylamide gel, 95—101
fractionation, 93—95
isolation, 82—84, 88—92
Chymotrypsin A, 122
Circadian periodicity, 152
Clostripain, 13
CM Sephadex® column chromatography, 115, 119
Collagen
biosynthesis, 165, 166
enzymatic digestion, 165
future age-related studies, 166
physical changes with age, 162, 163
Collagenase, bacterial, 165
Collagen cross-linking, 163, 164
Collagen isotypes, 163
Colloid endocytosis, 142
Competitive binding assays, 144
Concanavalin A, 146, 147
Concanavalin A-Sepharose chromatography, 146—151
Conformational heterogeneity, 138
Connective tissue, see also Collagen
complications in diabetics, 165
physical changes with age, 162, 163
Contamination, laboratory, 11
Coomassie® blue staining, 16, 17, 97,100, 101, 105
Corticosteroids, 78
Cross-reacting material (CRM), 126, 128, 130
Cross-reacting material ratio, 47, 48
Crystalline jack bean globulin, see Concanavalin A
(^{14}C) specific activity, 67
Cultured cells, see specific applications and cells
Culture medium, 57, 78
Cyanogen bromide, 12, 13, 25, 123, 125
Cyanomethyldithiobenzoate, 24
Cyclohexadione, 13
Cysteine misincorporation, 43, 44, 102
Cytochemical bioassay (CBA), 143

D

DABITC, see N,N-Dimethylaminoazobenzene 4′-isothiocyanate
DABTH, see N,N-Dimethylaminoazobenzene-thiohydantoins
DABTZ, see N,N,-Dimethylaminoazobenzene-thiazolinones
Dansyl chloride, 22, 23
Daunomycin-CH-Sepharose, 48
DEAE cellulose chromatography, see also Ion exchange chromatography, 93, 115, 116, 119, 125, 126
Decay constants, choice of, 64
Decay, simple exponential, 61
Degradation constants, see also Protein degradation, 67—69
Degradation rate, see also Protein degradation, 73, 74
Dehistonization of chromatin, 89
Densitometric scanning, 98
Deoxyribonucleic acid (DNA), see also Deoxyribonucleic polymerase; Ribonucleic acid
affinity techniques and, 93
coding mutations, 36, 37, 47
effect on protein solubility, 98, 99
mistakes in transcription and translation, 136
precursors, 52
RNA separation, 85
shearing, 83
Deoxyribonucleic polmerase, 50
Destaining of gels, 98
Dextran gel filtration, 114
Diabetic connective tissue, 165
Diamonstone method, 129
Diisopropyl fluorophosphate, 84, 122
Diisothiocyanate coupling, 25
N,N-Dimethylaminoazobenzene 4′-isothiocyanate (DABITC), 26
N,N,-Dimethylaminoazobenzenethiazolinones (DABTZ), 26
N,N,-Dimethylaminoazobenzenethiohydantoins (DABTH), 26
N-Dimethylaminopropyl-N-ethcarboniimide, 25
Dimethyl sulfoxide, 46, 52
DNA, see Deoxyribonucleic acid
Double immunodiffusion, 121
Double label technique, 65—70, 74
Dulbecco's phosphate buffered saline (dPBS), 147

E

Earls balanced salt solution, 86
Edman degradation, 13
Elastase, 122
Electrofocusing, see Isoelectric focusing
Electrophoresis, see specific types
Electrophoretic fractionation procedures, 95
Electrophoretic variants, 52
Enhance method, 105
Enolase, nematode, 6, 7
Enzymes, altered
age-related activity, 2
formation and accumulation, 136
formation hypothesis, 10
heat sensitivity, 4, 7
initial detection, 2—6
measurement by heat labile assay, 46, 47, 74
Ergot alkaloids, 78
Error catastrophe theory of aging, see error theory of aging
Error induction, models, 51, 52

Error measurement, see Protein, synthesis, error measurement
Error rate, 40—45, 49
Error theory of aging, 2, 36, 37, 49, 136
Erythrocytes
  G-6-PD activity, 119, 121
  TPI electrophoresis, 27
Ethionine, 45, 50—52
Ethylation, 45, 50, 52
N-Ethyl malemide, 51

F

Fibroblast, cultured
  applications in aging research, 76, 77, 83, 166
  error theory, 45, 48—50
  fetal lung, 73, 120
  protein degradation studies and, 71, 72
Fibroblast growth factor, 78
Fidelity of protein synthesis
  estimation, 36, 44, 45
  relationship with error rate, 49, 50, 136
Fischer 344 rats, 141, 142, 147
Flagellin, *E. coli*, 43
Fluorescamine, 21
Fluorography, 105
Follicle stimulating hormone (FSH), 137—139
Fractionation of chromosomal proteins, see also Chromosomal proteins, nonhistone, 84, 85
Fragmentation, primary, 12, 13
Frozen storage, effects on sampling, 6, 12
Fruit flies for error theory experimentation, 52
FSH, see Follicle stimulating hormone

G

Galactose, 139, 140
Gas chromatography, 24, 26
Gel electrophoresis, see specific electrophoresis systems
Gel filtration, see also specific gel filtration systems, 15, 16, 27
Gel filtration chromatography, 145
Gel fractionation and counting, see also Polyacrylamide gel electrophoresis, 105
Gel preparation, see also specific gel electrophoresis systems, 99, 101—103
Genome-associated proteins, 82, 85
Glassware cleanliness, 11, 96
Glucagon, 78
Glucosephosphate isomerase, 12, 13
Glucose-6-phosphate dehydrogenase (G-6-PD)
  age-related change and, 3
  apoenzyme, 124
  heat stable variant, 46
  hyperanodic form, 119, 121, 122
    heat lability, 120
    leukemic factor, 117, 122
    NADP, 123, 124
  immunological techniques, 116
  modification conditions, 117, 118
  post-translational modifications, 114
  properties, 115, 116
  purification techniques, 115
  thermolabile, 52
Glutamic acid, 41
Glutamic dehydrogenase, 2
Glutamine quantitation, 23, 24
Glycine bond cleavage, 12, 13
Glycoproteins, see Hormones, glycoprotein
Glycosylation
  effects on aging, 154 ,155
  hormone heterogeneity, 151—153
  polypeptide polymorphism and, 138—140
β-Glycuronidase, 3
G-6-PD, see Glucose-6-phosphate dehydrogenase
Growth hormone (GH), 137

H

H35 cells, 74
Half-life, protein
  heat labile enzymes, 74
  long-lived protein, 77
  measurement, 60—71
$^{3}H/^{14}C$ ratio, 66, 67, 70
HeLa cell chromatin proteins, 83, 85
Hemoglobin misincorporation, 37—43
Hexosaminidase A and B, 115
High-performance liquid chromatography (HPLC), 24, 26, 27
High-pressure liquid chromatography, 41
High resolution gel electrophoresis, see also Sodium dodecyl-sulfate-acrylamide gel electrophoresis, 49
High resolution X-ray information, 28
High sensitivity amino acid analysis, 10, 11
Histidine, 23, 48, 58
Histidino-hydroxymerodesmosine, 165
Histone(s)
  analysis, 101—106
  chromosomal proteins and, 82, 83, 89
  fidelity, 42
  isolation, 86—88
(^{3}H) label, 74, 104—106
Homology peptide mapping, 17, 20, 21, 27
Hormones, glycoprotein, see also specific hormones
  affinity chromatography, 146, 147
  bioassays, 143
  biological activity, 140
  carbohydrate composition, see also Glycosylation, 138, 154
  heterogeneity, 137, 138, 141
  pituitary, 136, 137, 140, 141, 144
  polypeptide polymorphisms, 136, 139, 154, 155
  protein degradation, see also Protein degradation, 78
  separation, 146
  serum levels in aged rats, 142, 143

HPLC, see High-performance liquid chromatography
Human chorionic gonadotropin, 137—139
Hyaline cartilage, 163
Hydrolysis tubes, 22
Hydrothermal isometric tension, 162
Hydroxyapatite, 99
Hydroxylamine cleavage, 12
Hydroxylysinonorleucine, 164
Hydroxyproline, 163—165
Hyperanodic forms of G-6-PD see Glucose-6-phosphate dehydrogenase, hyperanodic form
Hypophysectomy, 140

I

IEF/SDS system, see Nonequilibrium pH gradient gel electrophoresis, SDS-PAGE system
($^{125}I/^{131}I$) ratio, 75
Immunodiffusion techniques, 6, 115, 128
Immunological/biological potencies, 138
Immunological techniques, see also specific techniques, 114, 116, 126, 128, 129
Immunotitration, 2, 3
3-(3-Indolyl)propionic acid, 21, 22
Insulin, 78, 155
Intermittent perfusion, 76
In vitro systems, 57
In vivo systems, 57
Ion exchange chromatography, see also DEAE cellulose chromatography, 89, 93, 163, 164
Isocitrate lyase, altered
  isoelectric focusing, 6
  isozymes, 7
  nematode, 2, 4
Isoelectric focusing
  aldolase, 126, 128
  in detecting altered proteins and, 6, 7, 114
  electrophoresis and, 48, 49
  gel electrophoresis, 98—101
  tyrosine amino transferase, 129
  urea and, 15, 16
Isoelectric point modifications, 119, 120
Isoleucine
  microsequencing and, 26
  misincorporation, 37—41, 46
  substitution frequency, 43
Isotope reutilization, 68
Isozymes, 7

K

Ketoamine-linked glucose residues, 165
Kinetics of radioactive labeling, see Labeling, radioactive, kinetics
Km values, 6, 7

L

Labeling, radioactive, see also specific labeled substances
  kinetics, 63—70
  length of period, 70, 71
  mathematical aspects, 59—62
  medium removal, 71
Lactic dehydrogenase, 2, 3
Lectins, 146
Leucine
  calculation of protein degradation and, 70—72
  cell structure alteration and, 121
  microsequencing and, 26
  radioactive, 57, 58, 63, 66, 67, 70, 71
Leucine-free medium, 76
Leukemia, mononuclear cell, 142
Leukemic factor, 117, 118, 122, 123
Leukemic leukocyte extract, 119
Leukocyte enzymes, 121
LH, see Luteinizing hormone
Ligases, 44, 45
Liver
  amino acid release studies, 72
  mouse and rat in nuclear isolation technique, 83, 84
  regeneration, 58
Liver cells, human in culture, 46
Liver extracts, preparation, 129
Liver proteins, labeling of, 57
Luteinizing hormone (LH), 137—141
Lymphocytes and G-6-PD activity, 119, 120
Lysine, 23, 25, 43, 121
Lysosomal proteases, 6, 56
Lysyl residues, 163, 164

M

Malic dehydrogenase, 2
Mannose residues, 139
α-Mannosidase, 3
Marker enzymes, chromatin, 85
McKenzie bioassay, 143
Mercaptoacetic acid (thioglycolic), 21, 22
3-N-Mercaptoethane sulfonic acid, 22
Mercaptoethanol, 25, 51, 96, 97, 99, 100
Metabolic steady-state, 60, 67, 70
Methionine
  fidelity measurement and, 41, 42, 102
  residues, 102, 164
Methionyl bonds, 12
Methylation, 50, 52
N,N′-Methylenebisacrylamide, 96, 97
Methyl transfer reactions, 42, 45
5-Methyltryptophane, 51
Michaelis constant, 123, 128
Microheterogeneity, 138
Microsequencing
  general methods, 10, 24, 25

manual methods, 26
Mischarging of transfer RNA, 36, 44
Misincorporation, 50
Mispairing, 44
Model, laboratory experimental, see also specific models
aging process and, 76
animal, 141, 142
bacterial, 36
Molecular sieve chromatography, 39, 43, 88
Molecular specificity, 49, 50
Molecular weight analysis techniques, 114
Mutation mechanisms, 36, 37, 52

N

NADP, see Nicotinamide-adenine dinucleotide phosphate
NADPH, see Nicotinamide-adenine dinucleotide phosphate, reduced
Neuraminidase, 122, 141
Nicotinamide-adenine dinucleotide phosphate (NADP)
modifying proteins, 123, 124
purification of G-6-PD and, 115, 118
reduced (NADPH), 118, 123
Ninhydrin, 11, 21, 25, 164
2-Nitro-5-thiocyanobenzoic acid (NTCB), 13
Nonequilibrium pH gradient gel electrophoresis (NEPHGE), SDS-PAGE system, 101, 104
Nonhistone chromosomal proteins, see Chromosomal proteins, nonhistone
Nonidet 9-40, 96, 99
Nonionic detergents, 83, 85, 98, 99
NTCB, see 2-Nitro-5-thiocyanobenzoic acid
Nuclear isolation methods, 82—85
Nucleic acids, see also Deoxyribonucleic acid; Ribonucleic acid, 85

O

Oligosaccharide structure, 139
One-dimensional gel electrophoretic systems, 101—104
Ouchterlony double diffusion plates, 116

P

Papain, 122
Parafluorophenyl alanine, 51
Passage cells, 76—78
Pepsin, 122
Peptide, see also Polypeptide
alignment, 17
homology mapping, 17
hydrolysis, 21, 22
recovery from SDS-polyacrylamide gels, 16, 17
separation, 13—16
sequencing, see Microsequencing
staining and visualization, 16, 17
Peptide analysis, see also Peptide mapping
extraction from thin layer plates, 11, 17, 20
glassware cleaning procedure, 11
Peptide mapping
electrophoresis, 20
fragmentation into small peptides, 17, 20
staining and visualization, 21
recovery, 21
Phase contrast microscopy, 83
Phenol, 25
Phenylalanine, 57, 58, 70
p-Phenylenediisothiocyanate, 25
Phenylisothiocyanate (PITC), 24—26
Phenylmethylsulfonyl fluoride, 12, 84
Phenylthiohydantoin, 24, 26
Phosphoadenosine diphosphoribose (PADPR), 124
Phosphogluconate dehydrogenase, 115
Phosphoglucose isomerase, 115
Phosphoglycerate kinase, 6, 7
Phosphorylase, 12
o-Phthaldialdehyde, 21
Phytohemagglutinins, 146
PITC, see Phenylisothiocyanate
Pituitary chromophobe adenomas in rats, 142
Pituitary gland, 140, 149
^{32}P-labeled proteins, 106
Platelets, 121
Polyacrylamide gel electrophoresis, see also Sodium dodecyl-sulfate-acrylamide gel electrophoresis
analysis of chromosomal proteins, 95
gradient gel, 117
slab, 114
separation of peptides, 6, 15—17
Polyacrylamide gel plates washing procedure, 96
Polymorphism, see also Hormone, glycoprotein, polypeptide polymorphism; Thyrotropin polymorphism, 136—140
Post-transcriptional modifications, 50, 51
Post-translational modifications
aldolase, 114
amino acid analogues, 50, 51, 58
collagen, 165
glucose-6-phosphate dehydrogenase, 114, 119
lysyl residues, 164
polypeptide hormones, 136
PPO/DMSO method, 105
Precursor pool, amino acid, see also specific amino acids, 60, 61, 62, 71, 72
Proinsulin, 155
Proinsulin-like material (PLM), 141
Prolactin, 139, 144
Proline analogues, 52
Pronase, 115
Pro-opiocortin, 155
Prostaglandin E, 78
Protease inhibitors, 12, 83, 84
Proteases, 11—13, 28

Protein(s), see also Protein degradation; specific proteins
alteration, 2, 6, 7
detection methods, 36
post-translational, 114
bond cleavage with hydroxylamine, 12, 13
chromosomal, see Chromosomal proteins, nonhistone; Histones
hepatic, 58, 59
homology structural analysis, 12
hormones, see Hormones, glycoprotein
long lived, 77
physical properties, 56
proteolytic modification, see Proteases
sequencing, see microsequencing
synthesis, 57, 58
error measurement in, 36, 37, 41, 42
fidelity, 136
rates, 42, 48, 75
Protein degradation (protein turnover)
avoidance in chromosomal protein fractionation, 85
calculation, 72—74
effect of omitting amino acids, 76, 77
half-life, 60—71
HTC membrane and, 74—75
mathematical aspects, 59—62
measurement of, 56, 57
in cultured cells, 70—78
in vivo, 59—70
regulatory enzymes, 52, 56
role of nutritional status, 78
relation to age, 76—78
turnover time, 58, 60, 61, 66—70
Proteolysis, 6
PTH, see Phenylthiohydantoin
Pulse labeling, 58
Purification
altered enzymes, 3
protein in amino acid misincorporation, 38—43
Purity of labeled compounds, 40, 45
Puromycin, 71
Pyridine, 25
Pyridinoline, 164
Pyruvate kinase, 3

R

Radioactive precursors
error measurement and, 38—41
prevention of reutilization, 71
reutilization, 57, 58, 62
Radioactive tracers, see also specific tracers
choice of, 58, 70
measuring protein degradation, 56, 57, 62
Radioactivity, see also Labeling, radioactive; Radioactive tracers
flow of, 60
monitoring of release, 72—74
Radioimmunoassay (RIA)
conformational heterogeneity and, 138
TSH, 143—145
Radiolabeled proteins, analysis of
autoradiography, 105, 106
gel fractionation and counting, 105
general considerations, 104
fluorography, 105
Radiolabeling techniques in microsequencing, 24, 25
Radioreceptor assay (RRA), 144
Raney nickel, 13
Rats, aged stock colony, 141, 142
Reagent purity in microanalytical analyses, 11
Renal disease
aged rats and, 142
PLM levels, 141
Resins, see also specific resins, 25
Reticulocytes, 119, 120
Ribonucleic acid (RNA), see also Deoxyribonucleic acid
DNA separation, 85
error and, 36
precursor base analogues, 52
ribosomal, 45, 50
transfer (tRNA)
anticodon, 48
charging, 50, 57, 71, 72
mischarging, 44, 36
viral translation, 136
Ribonucleic acid polymerase, exogenous, 83
Ribosomes, fidelity of, 49, 50

S

Salt-urea extraction method, 89—93, 99
Sample overlay solution, 100
Scale fractionation, 95
SDS, see Sodium dodecyl-sulfate
Selective degradation of analogues, 45
Selective extraction procedures, 89
Sensitivity index (SI), 69
Sephadex columns, see also specific columns and procedures, 93—95
Sephadex G25 column chromatography, 117, 122
Sephadex G-200 column chromatography, 145, 148, 200
Sepharose, 93, 95, 147
SH groups, 6, 51
SH/ss ratio, 51
Sialic acid, 139—141
Single polyamide TLC identification, 26
^{35}S-labeled proteins, see also specific proteins, 105, 106
Slab electrophoresis, see also Sodium dodecyl-sulfate acrylamide gel electrophoresis, 95—97
S_1 nuclease digestion, 99
Sodium bisulfite, 84
Sodium borohydride, tritiated, 164
Sodium dodecyl-sulfate, 15, 96, 121
Sodium dodecyl-sulfate-acrylamide gel(s), 16, 17

Sodium dodecyl-sulfate-acrylamide gel electrophoresis (SDS), see also Polyacrylamide gel electrophoresis
- analysis of histones, 102—104
- chromosomal proteins, 93—98
- glucose-6-phosphate dehydrogenase, 117, 121
- high resolution, 49
- isoelectric focusing, 98—101

Solid phase sequencing, see also Microsequencing, 24, 25
Somatic mutation theory, 36
Specific activity
- enzymatic, 2, 3, 7
- misincorporated amino acids, 43
- proteins, 60—63, 66, 67
- ratio, 39

Spectral properties of altered enzymes, 5, 7
SS bonds, 51
Staining, see also specific stains
- aldolase, 126
- gels, 98, 100, 101
- glucose-6-phosphate dehydrogenase, 117

Staphylococcal protease, 13
Starch gel electrophoresis, 116, 126
Starvation, amino acid, 48
Steroids, 139
Streptomycin, 43, 48, 51
Subtilisin, 13
Sucrose density gradient centrifugation, 93, 114
Sulfhydryl content of proteins, 13
Superoxide dismutase, 6

T

TAT, see Tyrosine aminotransferase
Testosterone, 139, 141
Thin-layer chromatography, 24, 26
Thioacetylthioglycolic acid, 24
Thiozolinone derivatives, 25, 26
Threonine, 25
Thyroid stimulating hormone (TSH, thyrotropin), see also Thyrotropin polymorphism, 137—139
Thyrotropin (TSH) polymorphism
- assays, 143—145
- gel filtration chromatography, 145, 146
- investigational schemes, 147, 148
- rats and, 149—153

Thyroxine, 140
Tissue culture cells, see also specific cell lines, 83
3-N-p-Toluene sulfonic acid, 22
L-1-Tosylamide-2-phenyl-ethyl-chloromethyl-ketone, 84
TPI, see Triosephosphate isomerase
Triethylenetetramine resin (TETA), 25
Trifluoroacetic acid, anhydrous, 25
Triosephosphate isomerase, 2, 10, 12, 17, 27, 28
Tritium labeling in misincorporation method, 38
Triton-acid-urea gel electrophoresis, 102
Triton X-100, 83, 85, 102
Trypsin
- cleavage of arginyl bonds, 12, 13
- digestion, 17, 27, 165
- glucose-6-phosphate-dehydrogenase modifying factors and, 122
- stability to proteolysis, 115
- tissue culture cells and, 83

Tryptophan, 22
Turbatrix aceti
- enolase, 4, 5
- inactive enolase, 6
- isocitrate lyase isozymes, 2, 3, 7

Turnover, protein, see Protein degradation and turnover
Two-dimensional gel electrophoretic systems, 17, 48, 49, 98, 100, 101, 104
Tyrosine aminotransferase (TAT), 114, 129, 130

U

Urea
- gel electrophoresis and, 96,98, 102
- peptide separation and, 15, 16

Urea cycle, blocking of, 52, 58, 59

V

Valine, 58, 70
Valine free medium, 76, 77

W

Wobble error and substitution, 48, 52